Liquid Assets

Liquid Assets

*An Economic Approach for
Water Management and Conflict Resolution
in the Middle East and Beyond*

Franklin M. Fisher
Annette Huber-Lee
Ilan Amir
Shaul Arlosoroff
Zvi Eckstein
Munther J. Haddadin
Salem Ghazi Hamati
Ammar M. Jarrar
Anan F. Jayyousi
Uri Shamir
Hans Wesseling

With special contributions by
Amer Z. Salman
Emad K. Al-Karablieh

RESOURCES FOR THE FUTURE
WASHINGTON, DC USA

Printed in the United States of America

An RFF Press book
Published by Resources for the Future
1616 P Street, NW
Washington, DC 20036–1400
USA
www.rffpress.org

Library of Congress Cataloging-in-Publication Data
Fisher, Franklin M.
  Liquid assets : an economic approach for water management and conflict resolution in the Middle east and beyond / by Franklin M. Fisher, Annette Huber-Lee, Ilan Amir.
    p.   cm.
  ISBN 1-933115-08-4 (hardcover : alk. paper) — ISBN 1-933115-09-2 (pbk. : alk. paper)
  1. Water resources development—Middle East.   2. Water resources development—Jordan River Valley.   3. Water-supply—Economic aspects—Middle East.   4. Water-supply—Political aspects—Middle East.   5. Water use—Middle East.   I. Huber-Lee, Annette.   II. Amir, Ilan.   III. Title.
  HD1698.M628F57   2005
  333.91'00956—dc22   2004029563

The paper in this book meets the guidelines for permanence and durability of the Committee on Production Guidelines for Book Longevity of the Council on Library Resources. This book was designed and typeset in ITC New Baskerville by Kulamer Publishing Services. It was copyedited by Sally Atwater. Cover photo, "Landscape Seen Through Arch, Morocco," © Royalty-Free/Corbis. Cover design by Maggie Powell.

ISBN 1-933115-08-4 (cloth)   ISBN 1-933115-09-2 (paper)

About Resources for the Future *and* RFF Press

Resources for the Future (RFF) improves environmental and natural resource policymaking worldwide through independent social science research of the highest caliber. Founded in 1952, RFF pioneered the application of economics as a tool for developing more effective policy about the use and conservation of natural resources. Its scholars continue to employ social science methods to analyze critical issues concerning pollution control, energy policy, land and water use, hazardous waste, climate change, biodiversity, and the environmental challenges of developing countries.

RFF Press supports the mission of RFF by publishing book-length works that present a broad range of approaches to the study of natural resources and the environment. Its authors and editors include RFF staff, researchers from the larger academic and policy communities, and journalists. Audiences for publications by RFF Press include all of the participants in the policymaking process—scholars, the media, advocacy groups, NGOs, professionals in business and government, and the public.

Contents

About the Authors

Franklin M. Fisher is the Jane Berkowitz Carlton and Dennis William Carlton Professor of Microeconomics, Emeritus, at the Massachusetts Institute of Technology, where he taught for 44 years. He is also the chair of the Water Economics Project. He is the author or coauthor of 16 books, including *The Identification Problem in Econometrics; Folded, Spindled, and Mutilated: Economic Analysis and U.S. v. IBM;* and *Disequilibrium Foundations of Equilibrium Economics.*

Annette Huber-Lee is the director of the Water Programme for the Stockholm Environment Institute and senior research director of the Tellus Institute. Her research focuses on the integration of economic, engineering, and ecological approaches to solve environmental and social problems in a comprehensive and sustainable manner, and the development of innovative approaches to environmental policy and natural resources conflict management. She has coauthored papers in *Water Resources Research* and *Water International,* as well as chapters in books and other publications.

Ilan Amir is now retired as an associate professor from the Department of Agricultural Engineering at the Technion-Israel Institute of Technology, where he taught for 30 years. His research has focused on agricultural production systems, decisionmaking, and policy. He has also developed an irrigation machine and a foam system against frost. Among his publications are articles in the *Journal of Agricultural Systems* and the *Journal of Agricultural Engineering.*

Saul Arlosoroff is a director and chairman of the Finance/Economic Committee of the National Water Corporation of Israel. Previously, he managed the Water and Sanitation Program at the World Bank. He is a coauthor of *Conflict Management of Water Resources* and a contributor to *Between War and Peace; Water, Peace and the Middle East; Management of Shared Groundwater Resources;* and *Every Drop Counts.*

Zvi Eckstein is a professor of economics at Tel Aviv University and the University of Minnesota. He has published papers in the *American Economic Review,* the *Journal of Political Economy, Econometrica,* and the *Quarterly Journal of Economics.* Professor Eckstein, a fellow of the Econometric Society, was formerly head of the Eitan Berglas School of Economics at Tel Aviv University and is currently a coeditor of the *European Economic Review.*

Munther J. Haddadin is director of the Office for Integrated Development, Amman, Jordan, a courtesy professor at Oregon State University, and an affiliate professor at the University of Oklahoma. His book *Diplomacy on the Jordan: International Conflict and Negotiated Resolution,* addresses the conflict on the Jordan River system. He is coauthor of the forthcoming *An Inside Story of Peace Making: The Jordan-Israel Peace Negotiations* and is working on an edited book about water resources in Jordan. His articles have appeared in *Water International, Water Policy, International Negotiation, Geographical Journal,* and the *Brown Journal of World Affairs.*

Salem Ghazi Hamati has focused on computer science, hardware maintenance, and software development and has produced several computer models for application in various sectors of the economy and education. Mr. Hamati is proprietor and managing director of the firm IBS JO and is a managing partner of Jordan TechnoTrade L.L.C. His current postgraduate research is on the development of a computer model to incorporate artificial intelligence into computer-aided design software. He is coauthor of an article in *Water Resources Research.*

Ammar M. Jarrar is director of the Water and Environmental Studies Institute at An-Najah National University, Nablus, Palestine. His recent work has focused on challenges facing natural resources management and on effluent reuse. He has published several articles and reports in *AWRA Journal, An-Najah Newsletter,* and *Porous Media Flow.* He is the contributor to and author of several reports and studies in the field of water resources management and planning.

Anan F. Jayyousi is deputy director of House of Water and Environment and is an associate professor of water resources at An-Najah National University. His recent research has focused on the integration of decision support tools with hydrologic models. He recently served as a technical adviser to the Palestinian Water Authority negotiation team on water. He has several articles in local and international journals, such as the *Journal of Hydraulics, Porous Media Flow,* and the *An-Najah University Journal,* and recently contributed to a book titled *Water in Palestine.*

Uri Shamir is professor emeritus in the Faculty of Civil and Environmental Engineering and founding director of the Stephen and Nancy Grand Water Research Institute at the Technion-Israel Institute of Technology. He serves as president of the International Union of Geodesy and Geophysics and as a member of Israel's negotiation team on water with its neighbors. He has published papers in *Water Resources Research* and the *Journal of Water Resources Planning and Management.*

Hans Wesseling, formerly with WL|Delft Hydraulics, is a program manager in the Dutch Ministry of Public Works and Water Management. He has been involved in water resources management projects in Egypt, Yemen, Morocco, India, Vietnam, China, and elsewhere. He is a coauthor of a paper in *Water Resources Research.*

Emad K. Al-Karablieh is an assistant professor in the Department of Agricultural Economics at the University of Jordan. His research interests cover water economics, water allocation, risk analysis, and integrated water resources management. He has served as an environmental economics consultant to the Ministry of Water and Irrigation of the Jordan Valley Authority; private sector corporations; and international, regional, and local organizations. He has published several articles in *Agricultural Water Management* and *Agricultural Systems* and is a contributor to *Water Resource Policy and Related Issues in the Hashemite Kingdom of Jordan,* now under review by RFF.

Amer Z. Salman is an associate professor at the Department of Agricultural Economics and Agribusiness at the University of Jordan. He has conducted research and consultation in the economics of water allocation and water pricing. His articles have appeared in *Water International, Agricultural Water Management, Agricultural Systems,* and the *Quarterly Journal of International Agriculture.* He is a contributor to *Water Resource Policy and Related Issues in the Hashemite Kingdom of Jordan,* now under review by RFF.

"AND IT CAME TO PASS at that time, that Abimelech and Phicol the captain of his host spoke unto Abraham, saying: 'God is with thee in all that thou doest. Now therefore swear unto me here by God that thou wilt not deal falsely with me, nor with my son, nor with my son's son; but according to the kindness that I have done unto thee, thou shalt do unto me and to the land wherein thou hast sojourned.' And Abraham said: 'I will swear.'

And Abraham reproved Abimelech because of the well of water which Abimelech's servants had violently taken away. And Abimelech said: 'I know not who hath done this thing; neither didst thou tell me, neither yet heard I of it but today.' And Abraham took sheep and oxen and gave them to Abimelech; and they two made a covenant. And Abraham set seven ewe-lambs of the flock by themselves. And Abimelech said unto Abraham: 'What mean these seven ewe-lambs which thou hast set by themselves?' And he said: 'Verily, these seven ewe-lambs shalt thou take out of my hand, that it may be a witness unto me that I have digged this well.' Wherefore that place was called Beer-Sheba [Well of the Seven], because there they swore both of them.... And Abraham sojourned in the land of the Philistines many days."

Genesis 21: 22–34

—————

"A sign for them is the earth that is dead; We do give it life, and produce grain therefrom, of which ye do eat. And We produce therein orchards with date palms and vines, and We cause springs to gush therefrom."

Holy Qur'an, Surat Yassin 36: 33–34

"See ye the water which ye drink? Do ye bring it down from the cloud or do we? Were it Our Will We could make it salt [and unpalatable]; then why do ye not give thanks?"

Holy Qur'an, Surat Alwaqi'a (the Event) 56: 68–70

"Say: 'See ye?— If your stream be some morning lost [in the underground], who then can supply you with clear flowing water?'"

Holy Qur'an, Surat Al Muk (The Kingship) 67: 30

Preface

So important is water that there are repeated predictions of water as a *casus belli* all over the globe. For example, the UN-sponsored Third World Water Forum stated in August 2001 that water could cause as much conflict in this century as oil did in the last. Said Crown Prince Willem-Alexander of the Netherlands: "Water could become the new oil as a major source of conflict."[1] The next day, former Senator Paul Simon wrote,[2]

> Nations go to war over oil, but there are substitutes for oil. How much more intractable might wars be that are fought over water, an ever scarcer commodity for which there is no substitute?

He went on to say,

> Last year American intelligence agencies told President Bill Clinton, in a worldwide security forecast, that in 15 years there will be a shortage of water so severe that if steps are not taken soon for conservation and cooperation, there will be regional wars over it.

And those are but two very recent examples of many.[3]

Such forecasts of conflict, however, stem from a narrow way of thinking about water. Water is usually considered in terms of quantities only. Demands for water are projected, supplies estimated, and a balance struck. Where that balance shows a shortage, alarms are sounded and engineering or political solutions to secure additional sources are sought. Disputes over water are also generally thought of in this way. Two or more parties with claims to the same water sources are seen as playing a zero-sum game. The water that one party gets is simply not available to the others, so that one party's gain is seen as the other parties' loss. Water appears (as Simon says) to have no substitute, so it can only be traded for other water.

But there is another way of thinking about water problems and water disputes, a way that can lead to dispute resolution and optimal water management. That way involves thinking about the economics of water and shows, in fact, that water can be traded off for other things. To see that this is the case, consider the following example:

Conflicting water claims are part of the dispute between Israel and Palestine.[4] In particular, both lay claim to the water of the (rather loosely called) Mountain Aquifer (see Figure P.1[5]). Rain falls on the hills of the West Bank and becomes groundwater. Much of it flows west into pre-1967 Israel, where it has been pumped since before the State of Israel even existed (the wells being relatively shallow); some of it flows in other directions. Both parties claim the water under conflicting principles of international law.

Now, no matter how much you value water,[6] you cannot rationally value it by more than the cost of replacing it. Hence, the availability of seawater desalination places an upper bound on what water can be worth on the seacoast. In the case of Israel and Palestine, such desalination now costs roughly \$0.50–\$0.60 per cubic meter on the Mediterranean coast. Hence, water in the cities of that coast—Haifa, Tel Aviv, and Gaza, among others— can never be worth more than \$0.50–\$0.60 per cubic meter. (For convenience, we use the higher figure throughout the book.)

The water of the Mountain Aquifer, however, is not on the coast but rather underground and some distance inland. Such water has its own costs. The cost of extracting it and bringing it to the cities of the coast is (very roughly) \$0.40 per cubic meter. It follows that *ownership* of Mountain Aquifer water can never be worth more than about \$0.20 per cubic meter (\$0.60 − \$0.40).[7]

Now 100 million cubic meters (mcm) per year is a very large amount of water in the dispute, probably larger than the difference in the negotiating positions of the parties. It follows from what has been said, however, that 100 mcm per year of Mountain Aquifer water can never be worth more than roughly \$20 million per year. The conclusion would be the same if we used 200 mcm, 300 mcm, or more: *the sums involved are trivial between governments.* Israel's gross domestic product, for example, is approximately \$100 billion per year. The value of the water in dispute is not sufficient to obstruct a peace treaty, nor is it large enough to be worth a war.

The principal point of that example is not that desalination is the answer to water disputes. (Indeed, as shown later in this book, the desalination upper bound on the value of Mountain Aquifer water is too high and desalination itself not efficient in many situations.) The point is that thinking about water *values* rather than water *quantities* can lead to useful and surprising results.

The example itself is drawn (with updated values) from a statement made in 1990 by the late Gideon Fishelson of Tel Aviv University. In answer to a remark that the next war in the Middle East would be about water, Fishelson said,

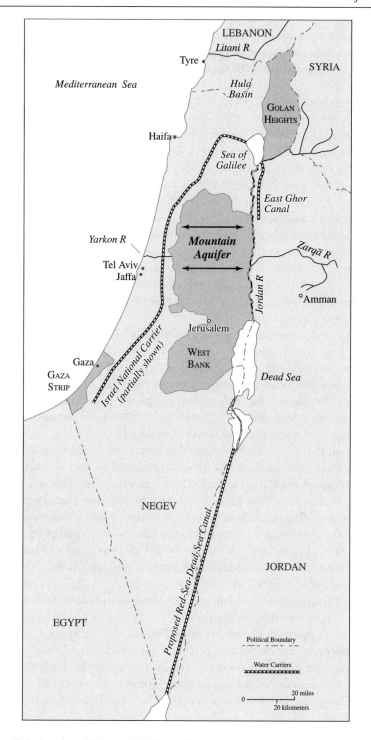

Figure P-1. Regional Map with Water Systems
Source: Adapted from Wolf (1994), p. 27.

Water is a scarce resource. Scarce resources have value. And the value of the water in dispute is bounded above by the replacement cost given by desalination, with that upper bound not very high.

Fishelson's statement provided the impetus for what is now named the Water Economics Project, or WEP, a project that has been in operation in various forms since 1992. The present volume describes the work of the project and its results. The applicability of that work is far from being limited to the Middle East, and accordingly, this book is about more than one topic.

Part I of the book (Chapters 1–4) is about methodology.

Chapter 1 continues the discussion of the value and economics of water and describes how that mode of thinking can lead to a powerful optimizing tool for water management and the cost–benefit analysis of infrastructure. It lays out the concepts that will be used throughout the book.

Chapter 2 describes the implementation of that analysis in a computer-driven tool, the Water Allocation System (WAS), and considers some of the implementation issues (e.g., the treatment of capital costs and of public policies toward water) in detail.

Chapter 3 discusses the Agricultural Submodel (AGSM), an optimizing model of crop choice, and how it is affected by water conditions and policies.

Chapter 4 returns to WAS and discusses what may be the most important concept: how WAS can be used in negotiations and especially in conflict resolution. We discuss the reasons that cooperation in water is likely to be superior to a standard division of water and how that can turn what appears to be a zero-sum game into a win–win situation.

Although Part I sometimes draws on examples from the Middle East, it is primarily methodological and of general applicability. Part II, on the other hand, presents in detail our results for Israel, Palestine, and Jordan. Although the base year used, 1995, is now a decade in the past and circumstances and situations have changed, the applicability of the methods has not, and the principal findings relate to years still in the future—2010 and 2020.

Chapter 5 applies both WAS and AGSM to Israel and considers such issues as the efficiency of desalination, various possible infrastructure projects, and water pricing policy.

Chapter 6 applies WAS to Palestine and considers both desalination and the effects of different water allocations.[8]

Chapter 7 applies WAS and AGSM to Jordan. We consider infrastructure issues, concentrating on various projects that could help alleviate the massive water crisis in Amman were no such projects to be undertaken. We find that the benefits of the projects considered (reduction of leakage, a pipeline from the Disi fossil aquifer, and the "Red-Dead" Canal) are very much interrelated.

Chapter 8 uses WAS to evaluate the benefits of cooperation in water along the lines developed in Chapter 4. We find those benefits quite sub-

stantial relative to the value of reasonably sized transfers of water ownership.

The overall purpose of this book is to promote a new way of thinking
about water, a way that leads to powerful methods of water management
and the analysis of water policy and infrastructure and that we hope will
break the impasse that too often characterizes water disputes.

A Brief History of the Middle East Water Project

We now turn to a description of the history of the Middle East Water Project, as the Water Economics Project was formerly known. That is necessarily
both an intellectual history and a political one.[9]

The Middle East Water Project is a joint project of Israeli, Jordanian,
Palestinian, American, and Dutch scholars. As described below, it has, since
1996, been facilitated and financed by the government of the Netherlands
with the consent (although not always the assent) of the three regional
governments.

The project officially began life in 1992 under the auspices of the (now
defunct) Institute for Social and Economic Policy in the Middle East,
directed by Leonard Hausman. The Institute was then located at the John
F. Kennedy School of Government at Harvard University. But the intellectual genesis of the Middle East Water Project occurred at a conference
the Institute held in London in 1990, when Gideon Fishelson made the
remark quoted above. That remark struck Franklin Fisher with great force,
and when the Institute decided in 1992 to have a project on water, Fisher
agreed to be the chair.

In fact, although the project was formally in existence in 1992, nothing much happened in that year. There was an abortive preliminary conference at which the participants were (perhaps not surprisingly) more
concerned with arguing over water ownership rights than with the value
of water, but the project was moribund until early fall 1993. In October
of that year, a second preliminary conference was held, and three things
had changed.

First, the Oslo agreement had been signed in September. As we were to
discover many times, our work went more smoothly when relations among
the regional parties were good than when they were bad. Second, a largely
different group of people were present at the second conference than had
attended the first. There was considerably more receptivity to the use of
economic analysis (broadly construed). Third, the centerpiece of the conference was the discussion of a preliminary model of the water economies
of Israel, the West Bank, and Gaza, created by Fishelson and Zvi Eckstein
with the assistance of Yuval Nachtom (all then at Tel Aviv University).[10] That
model was the precursor of the WAS model discussed in this book. There

was general agreement that the project would produce a more accurate and better model of the water economies of Israel, Palestine, and Jordan.

That agreement, however, was in large part due to the insistence of Ellen Paradise Fisher (in discussions with Franklin Fisher) that the project must find a way of handling the differences between social and private benefits (and costs) of water. As we shall see, that became a major component of the methodology, allowing users of the model to insert their own values and policies.

The ensuing work (in what we now think of as "Phase 0" of the project) went on from late 1993 into early 1996. The project operated in teams: three (nongovernmental) "country teams" and a "central team" in Cambridge, Massachusetts. Although the country teams were primarily responsible for data collection and model construction in their respective countries and the central team was primarily responsible for model-building theory and coordination, those responsibilities were often shared more widely than such a division indicates, and a good deal of interaction took place.

Now, the original Eckstein–Fishelson model ran in terms of the usual economic concepts of supply and demand. Since it (wrongly) assumed that all water conveyance ran from north to south (along the Israeli National Carrier), in any equilibrium, price in any district would differ from price in a district farther north by the cost of conveyance.[11] Hence, knowing the conveyance costs, one could find the equilibrium prices in this highly nonlinear model by iterating on a choice of price in the northernmost district.

This was computationally efficient but, as it turned out, very confining as to what could be accomplished. The late Robert Dorfman, in particular, quickly began to insist (and never stopped insisting) that the model be formulated in terms of an explicit maximization problem. Although it was true that equilibrium prices that equate supply and demand do lead to situations where net benefits (social welfare) are maximized (see Chapter 1), that maximization is only implicit and only deals with private benefits. Further, solving the maximization problem in this way would be efficient only if all water flows were truly unidirectional and in a linear conveyance system. Moreover, the opportunity to deal with more than one water quality—in particular, with recycled water—in this way was very limited.

When Fisher finally was wise enough to turn away from day-to-day problems and listen to Dorfman, the latter produced N. Harshadeep, then a doctoral candidate in applied engineering (and now at the World Bank). Harshadeep served as the project's principal programmer until 1996, introducing the use of GAMS into an explicit maximizing version of the model. The program now has expanded facilities and a somewhat different look and feel from the one he wrote, but Harshadeep's fingerprints are very much on it.

As Phase 0 of the project progressed, the high value of Dorfman's insistence on explicit maximization became apparent. The adoption of that suggestion allowed a number of advances:

- Most important was the ability to fully accommodate views of the special social value of water that did not coincide with private benefits. As explained in Chapters 1 and 2, this kept the WAS model from being a narrowly conceived economic model and permitted explicit incorporation of policies and constraints imposed by the model user.
- Explicit maximization brought the crucial role of shadow values to the fore.
- That treatment and the calculation of shadow values also made possible the analysis of what should happen to the prices charged consumers (if those are not fixed in advance) when there are environmental charges and, especially, when there are scarcity rents earned by recycling facilities.
- Further, explicit maximization made the calculation of benefits and costs from policies and infrastructure projects a straightforward affair, turning the model into a powerful tool for such analysis.
- At first, the role of the estimation of water values in conflict resolution seemed only the still important one of expressing conflicts as disputes over money, thus making them easier to resolve. But as we will see in Chapters 4 and 8, explicit maximization of benefits made it possible to calculate the value to each party of cooperation in water through trade in water permits—a major step forward.

Throughout Phase 0, the project members—usually, but certainly not exclusively, Fisher—met with political leaders in the region as well as in the United States. Those unofficial meetings took on a more official character toward the end of 1995.

With model construction nearing what then seemed completion (a view that now seems laughably naïve), Leonard Hausman arranged a meeting between Fisher and Nabil Sha'ath, then the Palestinian minister of Planning and International Cooperation, when Sha'ath visited Cambridge in late October 1995.

Minister Sha'ath informed Fisher and Hausman of the so-called "Dutch Initiative." At the second Middle East Economic Summit in Amman, Jordan, the Dutch government had stated its readiness to facilitate joint projects among Israel, Palestine, and Jordan, particularly projects involving regional infrastructure. Egypt had then expressed an interest and was added to the group.[12] This initiative was to be governed by the planning ministers of the regional parties, under the chairmanship of Jan Pronk, the Dutch minister for Development Cooperation within the Foreign Ministry. The three regional ministers involved (omitting Egypt) were Yossi Beilin of Israel,[13] Rima Khalaf of Jordan, and Nabil Sha'ath himself.

Sha'ath was very strongly of the view that the Middle East Water Project would be perfect for the Dutch Initiative and said that he would propose it. He urged Fisher to visit the other planning ministers and speak with them about the project before the expected ministerial meeting in the Hague at the end of January 1996, at which projects were expected to be considered and adopted. He further sent Fisher to see his own deputy minister, Ali Sha'at in Gaza.

The visits took place, first in November 1995, and then in January 1996, after which the Middle East Water Project was adopted by the meeting of ministers. Nevertheless, it was not until July 1997 that all contracts were signed and Phase 1 began full operation (although several trips to the region were supported by the Dutch government before that time). This delay was caused by two unrelated phenomena.

First, the Israeli government signed off on the project, together with a budget and lists of participants, in early May 1996. However, the same month brought the election of a new Israeli government, led by Benjamin Netanyahu. Even though the outgoing government had agreed to the project, the Dutch government then sought to obtain the Netanyahu government's explicit assent. Because water was (and is) such a politically important subject, it took until the end of 1996 to ask for and obtain that assent.

Then there followed six months of contracting arrangements involving Harvard, the three regional governments, and three private teams. Budgetary matters in the Dutch Foreign Ministry also arose. At the last moment, with all the teams ready, but when it appeared that further delay would be catastrophic in terms of holding things together, the Dutch Minister Pronk, as strongly urged by his staff, agreed to the start of Phase 1 as of the beginning of July 1997.

Phase 1 brought with it a number of significant changes in personnel. First, Harshadeep received his degree and left for employment at the World Bank. He was succeeded principally by Annette Huber, now Annette Huber-Lee, who has been the project's principal programmer ever since. But Huber-Lee has not merely been a programmer. She also worked with the teams in the region as well as the central team and very much contributed to the writing of this book.

Then, in January 1996, Ilan Amir brought to the project his ideas for an optimizing model of crop choice—what became the Agricultural Submodel discussed in Chapter 3 and extended in the appendixes to Chapters 5 and 7. He spent an 18-month sabbatical in Cambridge, starting in March 1996, working on AGSM and contributing to the general project.

Next, in May 1997, the Dutch Foreign Ministry retained the firm of Delft Hydraulics. Hans Wesseling, of that firm, was to act as a reporter to the ministry and also to assist the project, where possible. That second role became very large indeed, with Wesseling soon a major project participant. As did Huber-Lee, he contributed not only to model-building theory but also to working with the country teams on data collection and model use.

Finally, the country teams also changed. The structure became one in which each regional party had both a private team and a supervisory governmental steering committee. But the three parties were not all alike in the relative roles played by private and public participants. In Israel, the steering committee acted only in a general oversight capacity; in Palestine and Jordan, members of the steering committee were far more active.

In Israel, the leaders of the private team were Shaul Arlosoroff, Zvi Eckstein, and Uri Shamir.[14] They were joined by Amir and by Elisha Kally. Assistance beyond the call of duty was provided by Yossi Yakhin.

In Jordan, Munther Haddadin, first as private team leader and then as water minister, was in overall charge. Members of the team included Salem Hamati and, for agriculture, Emad Al-Karablieh and Amer Salman. (Their substantial contribution is reprinted in the Appendix to Chapter 7.) A very significant role was played by Hazim El-Naser, now water minister.

In Palestine, Anan Jayyousi and Ammar Jarrar of An-Najah University (the latter working largely on agriculture) became the central figures of the private team. From the government, substantial contributions were also made by Khairy Al-Jamal and Said Jalala, and by Ali Sha'at, who was in overall charge.

In June 1998, Harvard's Kennedy School began a process that was to dissolve the Institute for Social and Economic Policy in the Middle East by late fall. The reasons for that action had nothing to do with the project. Indeed, strong representations were made to Fisher as to the commitment of the Kennedy School to the project and the strong desire of the school to keep it there. Fisher agreed.

Those commitments, however, evaporated by the end of September, when—again for reasons having nothing to do with the substance of the project—the Kennedy School decided to drop it.[15] As a result, Harvard's name is no longer associated with the project. Management of the project was quickly shifted to Delft Hydraulics, where it remained. This was a very fortunate change.

Phase 1 of the project formally concluded in early 1999, and the results reported in this book are largely based on it. During that phase, each of the three regional parties worked on the construction of a model of its own water economy. That work was largely bilateral, with each regional team working with the central team. There were, however, occasional joint meetings and discussions on common topics and problems. As described in this book, the models were single-year, steady-state ones (although the conditions for the single year can be varied). Construction of a full multiyear model has begun but is not far advanced.

During Phase I, the project encountered rocky political waters. When, in 1998, Israel's then-Minister of Infrastructure Ariel Sharon and his water commissioner, Meir Ben Meir, learned of the project, they were not happy. Not fully understanding what the project could offer, they made a number of incorrect public statements about it. Nevertheless, they did not go so far as to have Israel withdraw.

Further, the attitude of Palestine toward the project also varied, but this had mostly to do with the state of relations between Palestine and Israel, which were usually poor during the Netanyahu years (although, sadly, not nearly so poor as they were later to become).

At the close of Phase 1, the Dutch government proposed an interim phase in which each party would seriously consider the adoption of the project's techniques into their own water planning agencies. (A further phase involving consideration of actual cooperation in water was to be a later possibility.) Eventually, all three parties agreed to such an interim phase, with the Israeli government assenting in spring 1999 (before the election of Ehud Barak), the Jordanian in the summer, and the Palestinian not until early autumn 2000. That delay involved various contracting problems.

Alas, immediately after the Palestinians gave their assent, the El-Aqsa Intifada began, and discussions of project continuation ceased. At present, given the deterioration of relations in the region, it is unclear as to whether or how the project will continue, although the Dutch government remains committed to it and there are still expressions of interest in the region.

Nevertheless, work goes on. We believe that we have developed an important way of thinking about water and a tool that is applicable to water management and water dispute resolution all over the world. We hope the reader will agree.

Previous Work

Of course, the realization that water can (and should) be treated as an economic commodity—with special properties—is not new. It goes back at least as far as the Harvard Water Program of the 1950s and 1960s (see Eckstein 1958; Maass et al. 1962; Hirshleifer et al. 1960). That realization has led to a large literature on water markets, to policy recommendations for water markets, particularly by the World Bank, and in some cases to the adoption of such recommendations (see, e.g., Saliba and Bush 1987).

But for reasons discussed in Chapter 1, actual water markets often do not lead to efficient or optimal results. There are several reasons, but perhaps the most important is that allocating water involves social benefits and costs that are not simply private benefits and costs. Although some of these externalities, such as environmental issues, are recognized in the literature, others are not. In particular, the subsidization of water for a particular use (usually agriculture) implies a policy decision that there are social benefits from such use above and beyond those realized by the direct users themselves. But private markets fail to handle this, and where such views are held by policymakers, such failure properly makes them unwilling to institute actual market mechanisms.

Our work, as will be seen, overcomes such difficulties: the use of an explicit optimizing model allows the policymaker to impose his or her own views about water values or, equivalently, to impose certain water policies and also to take other externalities into account. This is a form of systems analysis.

The application of systems analysis to water management is also not new; see, for example, Rogers and Fiering (1986) and the papers cited below. As Rogers and Fiering (1986, 146S, 150S) state,

> Systems analysis is particularly promising when scarce resources must be used effectively. Resource allocation problems are worldwide and affect the developed and the developing countries that today must make efficient use of their resources. At any given time some factor such as skilled manpower, energy, transport, material or capital is in short supply and that scarcity impedes progress. Under these conditions, governments and their planners, but particularly those in less developed countries (LDC's), are constantly faced with the need to make the best use of those resources that will be needed at further stages of progress, and with the future need to balance current needs against investment for the future. This is the domain of systems analysis.

However, they also point out that

> ...many, if not most, conventional uses of mathematical programming to solve water resource problems pursue optimality indiscriminately....Unfortunately, the stated...goals and measures of merit do not fully reflect the true concerns of the decision makers, and experience shows that mathematical programming, while often used in some vague way by planners, is rarely used to make the critical decisions associated with project planning.

We attempt to solve this problem by permitting the user of the Water Allocation System model to impose his or her own values.

In addition, there are other differences between our work and its predecessors. First, much previous work concerns the cost–benefit analysis of particular projects, whereas we provide a tool for the overall management of water and water projects, taking into account countrywide (or regionwide) effects while using a model that is geographically disaggregated. Second, perhaps the most familiar use of such models is that of minimizing cost in connection with fixed demand quantities.[16] The WAS model goes beyond that in more than one respect:

- WAS takes account of demand considerations and the benefits to be derived from water use rather than fixing water quantities to be delivered.
- WAS permits the user to impose social values that differ from private ones and to impose policies that the optimization must respect.

- Beyond mere optimization, we show how such models can be used in conflict resolution, an area highly important in water issues. We show that such models can be used to value disputed water, thus effectively monetizing and deemotionalizing the dispute. Moreover, for international disputes, we show how one can analyze the water systems of each party separately, testing options of links to other parties, or analyzing the combined territory of two or more parties as one. This provides estimates of the benefits of cooperation, which can then be weighed against the political issues involved in such cooperation.

Nevertheless, as the first epigraph to this book shows, the principal idea—that water can be traded off for other things in the context of a peaceful resolution of differences—is roughly 3000 years old. It is found in the story of the patriarch Abraham, revered by Jews, Muslims, and Christians, and Abimelech, a king of the Philistines. Abraham and Abimelech did not know about computers or modern economics or optimizing models. But they understood a principal message of this book.

Acknowledgments

In addition to those already mentioned in the history of the project, many others deserve not only mention but gratitude for their contributions, advice, and support. In addition, some of those who were mentioned deserve a fuller statement of thanks than was given in the history. We apologize to anyone we may have inadvertently omitted.

We give these acknowledgments more or less in chronological order, but there is at least one person whose assistance ran through the entire project. That person is Leonard Hausman, the former director of the Institute for Social and Economic Policy in the Middle East. He provided funds for Phase 0, but beyond that provided (and continues to provide) contacts, advice, and enthusiasm. Without him there would have been no project.

Phase 0 (1993–1996)

Central team. During this period, the central team was principally composed of Fisher, Eckstein (then visiting Boston University), Hillel Shuval of the Hebrew University (visiting the Institute), and the late Robert Dorfman of Harvard, consulting with the team. Theodore Panayotou of the (now defunct) Harvard Institute for International Development participated in both deliberations and visits to the region. Peter Rogers of Harvard also provided useful advice. Occasional visits by Atif Kubursi of McMaster University were also very helpful, especially later, when tutoring others in the use of the model became important.

The central team acquired two members as research assistants. But these "assistants" quickly became full participants in the project, providing more than simply support. Their names appear as coauthors on an earlier "Liquid Assets," a report circulated, but not published, in 1996.

The first was Aviv Nevo, now at the University of California, Berkeley. He acted as the principal assistant in economics. He interacted with the country teams on data and other questions and strongly participated in the central team's discussions of economic analysis.

The second "assistant" was N. Harshadeep, whose work was described above.

At the Institute for Social and Economic Policy in the Middle East, valuable administrative help and political advice was provided by Anna Karasik (now Anna Thurow), who has remained a friend of the project ever since.

Israel. The Israeli team consisted of Zvi Eckstein, the late Gideon Fishelson, and Hillel Shuval, assisted by Yuval Nachtom. Yehudah Bachmat contributed a steady-state model of the hydrology of the Mountain Aquifer. Following a (somewhat premature) conference in Cyprus in June 1994, Shaul Arlosoroff provided invaluable information, advice, and criticism.

Jordan. The Jordanian team consisted of Elias Salameh, University of Jordan, and Maher Abu-Taleb and Iyad Abu-Moghli of Environmental Resources Management Consultants. Although not part of the team during this period, Munther Haddadin and Jawad Anani gave, and were to continue to give, enormously helpful advice and support. Ali Ghezawi, who attended the October 1993 conference in Cambridge, was also helpful.

Palestine. During the early part of Phase 0, Jad Isaac of the Applied Research Institute in Bethlehem was the leader of the team and remained throughout that phase as a major force. He was later joined as coleader by Marwan Haddad of An-Najah University. Others on the team were Abdel Rahman Al-Tamimi, Mohammed Al-Turshan, Numan Mizyed, Yousef Naser, and especially, Taher Naser-ed-dine and Mustafa Nusseibeh. The Palestinian team operated through the Palestine Consultancy Group, Sari Nusseibeh, president, and Issa Khatar, director.

Phase 1 (1996–1998) and beyond

The Dutch government. Our greatest thanks go to the government of the Netherlands, which, motivated by conscience rather than self-interest, has continued to support the project through good and bad times and hopes even today to continue to do so. We are, of course, particularly grateful to Minister Jan Pronk and his successor, Minister Eveline Herfkens, but our thanks most certainly do not stop there.

Ambassador Como van Hellenberg Hubar in Tel Aviv was a constant source of wisdom and encouragement, and his assistants, Joanna van Vliet and Gert Kampman, were also very helpful, as was Michael Rentenaar and (later) Maarten Gieschler of the Dutch Representative's office in Ramallah.

In the Hague itself, the project found a number of committed supporters. In particular, we wish to thank Gerben de Jong, his successor, Norbert Brackhuis, and Bert Diphoorn, Angelique Eijpe, and Kees Smit Sibinga.

But there is one person in the Dutch Foreign Ministry for whom no expression of thanks can possibly be adequate. That is Louise Anten, who was in charge of the project until mid-1999. Not even all the authors of this volume can know how much we owe to her, but at least three of us, including the chair, can never forget her unswerving support and dedication to a cause that at times seemed hopelessly lost.

Outside the government, at Delft Hydraulics, Geert Prinsen assisted Wesseling and Huber-Lee in working with the country teams on data collection and model use. Jan Verkade provided efficient and appreciated administrative support.

Central team. As Phase 1 began in early 1996, Aviv Nevo and N. Harshadeep left the project. In the area of programming, Harshadeep was succeeded for a while by other students of Peter Rogers, including Trin Mitra and, as already mentioned, Annette Huber-Lee. Only Huber-Lee remained with the project for any substantial period.

Aviv Nevo was succeeded as the principal assistant in economics by M. Daniele Paserman, who also assisted with programming and remained for more than a year.

Howard Raiffa of Harvard provided advice and discussion on conflict resolution. Robert Mnookin of the Harvard Law School also helped here.

Of course, until 1998, the staff of the Institute for Social and Economic Policy in the Middle East often continued to provide valuable assistance.

Israel. After the meeting with Nabil Sha'ath and before the project was adopted by the Dutch Initiative, Fisher met with Alon Liel, then director-general of Yossi Beilin's ministry. He has been a tower of support and a source of wise advice ever since. We also thank his then assistant, Amir Tadmor, plus Rafi Bar-El, Roby Nathanson, and Yair Hirschfeld for their interest, assistance, and support.

The governmental steering committee was under the chairmanship of Dan Caterivas and included Yossi Dreizin, Mayor Admon, and Rafi Benvenisti.

There have also been meetings in Israel with people in and out of government. We wish particularly to thank Haim Ben-Shahar and Yoav Kislev for their advice and support.

Jordan. In addition to those named in the brief history given above, others in the Jordanian government who played a role were Faisal Hassan, Boulos

Kefaya, and Hani Mulki, who succeeded Haddadin as water minister. Hassan, in particular, gave us very substantial help and advice on writing the final version of Chapter 7, as did Ra'ed Daoud.

But we also owe a great debt to Jawad Anani, former deputy prime minister, foreign minister, and adviser to the Royal Court. He provided unstinting assistance and support, helping us over political obstacles.

Outside Jordan, we are grateful to Julia Neuberger for her support.

Palestine.[17] In addition to those listed in the brief history, Khalid Qahman was particularly active in the early part of Phase 1. Others who contributed include Naim Al-khatib, Husam Al-Najar, Basil Nasser, Ghassam Abu-Ju'ub, Wael Safi, Mohamed Al-Wali, and especially Luai Sha'at and Ihab Barghouti. Karen Assaf provided interest and advice.

Members of the Palestinian private team included Yousef Abu-Mayleh (head), Mohammed Abu-Jabal, Moustafa El-Baba, Belal Elayyan, Mahmouid Okasha, and Fayez Shaeen, as well as coauthors Anan Jayyousi and Ammar Jarrar. The Palestinian steering committee included Omar Draghmeh and Nizar Wahidi, as well as Khairy El-Jamal and Said Abu Jalala, who are mentioned in the history.

Palestinian participation in the project was under the leadership of Ali Sha'at, deputy minister of Planning and International Cooperation. His role extended beyond administration to active participation.

We also benefited from conversations with Nabil Sharif, the head of the Palestinian Water Authority.

And, as is obvious from the history, we are very grateful for the support of Minister Nabil Sha'ath.

United States. We are indebted to Don Reisman for editorial advice and assistance, to Brian Palmer for help with Excel, and to Theresa Benevento for general assistance and manuscript preparation. We thank A. Myrick Freeman III and three other (anonymous) reviewers for RFF Press for comments and advice. Over the past few years, the writing of the book has been supported by the research funds of the Jane Berkowitz Carlton and Dennis William Carlton Professorship of Microeconomics at MIT, for which we are very grateful. Finally, a special thank you goes to Ellen Fisher, who has lived with this project for more than a decade. As we all know, that has not always been an easy task.

FRANKLIN M. FISHER
Cambridge, Massachusetts

Part I

General Methodology

Water and Economics

Water is essential for human existence, and in many regions of the world, water is scarce. As a result, conflicting claims on existing water resources are often put forward. These are sometimes conflicting claims by different political entities—countries, states, or provinces—and sometimes conflicting claims by different types of water users—farmers, households, industry, environmental groups. The first type of conflict is often predicted to be a likely cause of future wars, but even the second is a very serious matter.

Claims to water are almost invariably stated in terms of water quantities. The contending parties put forward conflicting arguments as to their rights to certain water sources or certain amounts of water. Those arguments are based on various principles; the principles often conflict; and each party bases its claim on the principle that results in an outcome it considers favorable. Often there is much to be said for each side, and the resolution of the claims is not an easy matter even when the parties are otherwise friendly.

This book sets forth a wholly different way of looking at water disputes, water allocation, and water management. That way is based on economic principles, although it is by no means restricted to economics narrowly construed. To see why such an approach may be useful, consider the following.

Water is not generally scarce in terms of quantity. If for no other reason, the availability of seawater desalination means that there is abundant water for the world as a whole and for any country that has a seacoast. But of course, seawater desalination is expensive (current estimates run roughly 50 to 60 cents per cubic meter on the Mediterranean coast of Israel and Gaza), and conveyance facilities to locations far from the sea may themselves be expensive or even nonexistent. There are two lessons to be learned here. First, water scarcity is a matter of cost and value, not merely of quantity. Second, the value of water and also its scarcity will be different in different locations. These are two of the principal themes of this book.

Microeconomics[1] is basically about the allocation of scarce resources and about the relation of the value of those resources to their scarcity and their allocation. The fact that water is essential for human life makes water and its allocation very important, but it does not exempt it from the principles of microeconomics.

For many people, of course, this seems unnatural. If water is essential for human life, why isn't it beyond price? How can one put a value on something so important? Sometimes, indeed, one encounters the opposite view (occasionally from the same people): Water is a natural right. It should be charged for at the direct cost (extraction, treatment, conveyance, etc.) of providing it.

Both those views are wrong, and it is instructive to see why that is so. An important (but not basic) reason that water is not somehow "beyond price" has already been given. No matter how important water is and no matter what special values are believed to attach to water in certain uses, it is irrational to value water at more than the cost of replacing it.[2] Hence the possibility of seawater desalination places an upper bound on the value of water.

More than replacement costs are at issue here, however; demand factors also play a role. To see this, observe that if water were truly beyond price, then it would pay any country with a seacoast to desalinate seawater and convey the resulting sweet water to any location no matter how far. Landlocked countries would make any offer, no matter how large, to have desalinated water delivered to them. Obviously, this does not happen because the costs involved are too high. But then the value of water at particular locations cannot exceed those costs.

On the other hand, the value of water does not merely consist of direct costs, such as extraction, treatment, and conveyance. Consider a lake and suppose, for simplicity, that it costs nothing to take water from the lake in terms of direct costs. Suppose, however, that there are sufficient people living near the lake that, at zero cost, their total annual demand for water is greater than the available supply (say, the annual renewable amount in the lake). Then the value of the water in the lake is not zero; if there were additional water, there would be positive benefits to the people living on the lakeshore and they would be willing to pay for those benefits.

This example is important, and we shall return to it. For the present, observe the following:

- The fact that the value of the water in the lake is greater than the direct costs of supply means that the lake water has a positive *scarcity rent* (a concept that will be rigorously defined later).

- The scarcity rent is a measure of scarcity. Indeed, with only a few people living near the lake, the same water, equally essential for human life, might not be scarce at all. In that case, its scarcity rent would be zero.
- It is the scarcity of water and not directly its importance for human life that makes water valuable. Water is valuable only where it is scarce.

As the above discussion suggests, our analysis will run in terms of values rather than quantities. To do this, however, we must first dispose of another common misconception.

The person, group, or political entity that owns a particular amount of water and uses the water does not generally obtain it for nothing. If someone else values the same water and would pay for it, then the using owner is giving up the value for which it could sell it. In this case, the owner is said to incur an *opportunity cost*.

Now, the behavior of the owner in such circumstances is worth analyzing. Consider the owner's decision about whether to consume the last cubic meter of water owned. If the value of that cubic meter of water to the owner is greater than the opportunity cost (the price at which it could be sold to another), then the owner will consume the water itself. On the other hand, if the opportunity cost of the last cubic meter is greater than its value to the owner, then the owner will gain more by not consuming it. But these considerations are the same as those in the mind of any buyer: If the water to be purchased is worth more than its price, then it is worth purchasing; if the water to be purchased is worth less than its price, then it will not be purchased. In effect, a water owner that uses the water itself is purchasing the water; it is purchasing the water from itself. The difference between owning the water and not owning it is only a matter of where the money goes.

This, in turn, reveals an important set of propositions:

- The ownership of water is the ownership of the money value that the water represents.
- Who owns water and who uses water are not the same question. Although they are both important questions, they are analytically independent.[3]

It is important to understand that the methods put forth in this book do not deal directly with the question of water ownership. Rather they deal with the question of optimal water *usage*—of providing tools for the efficient management of water systems for the public good, taking into account the values of the tool user. On the other hand, by focusing on the costs, values, and scarcity rents of water, our methods do provide an estimate of what water ownership is worth and of the benefits to be had from cooperation rather than conflict in water. In that sense, our analysis may make conflict resolution easier by providing a new mode for thinking about water.

We shall have more to say about such wider issues later on, but our main focus for the present is on optimal water management for a particular political entity, say, a country. Of course, even that discussion involves conflict resolution, since it implicitly or explicitly involves the resolution of conflicting claims to water by various user groups.

1. Markets and the Management of Natural Resources

If economists had only one thing to say to the rest of the world, it would concern the way in which private, competitive markets lead to an efficient and indeed, in some sense, optimal allocation of resources. The theorems that describe this—the First and Second Welfare Theorems—are doubtless the most important results of microeconomic analysis. But if economists had a second thing to say, it probably would (or at least should) concern the circumstances under which the two welfare theorems do not apply and markets do not lead to an optimal result. It is important to understand that such is the case for water (for reasons given below) and that we are definitely not recommending a system of private water markets. Nevertheless, the role of markets and the reasons for their failure are central in our analysis. Hence we begin with a brief discussion of the issues involved in the management of natural resources through markets.

Consider a fictitious mineral, Ozite, which is nonrenewable. Society wishes to have Ozite extracted optimally. Since extracting more Ozite now means that there will be less to be extracted later, society's interest includes consideration of the rate of extraction.[4] Further, Ozite, when used, will bring benefits, and the extraction of Ozite will involve costs. As a general matter, both benefits and costs in a given year will depend on the amount extracted in that year—costs because extracting more Ozite in a given period will typically cost more, and benefits because additional Ozite will presumably be useful. Note, however, that it is quite possible, even likely, that marginal benefits (the benefits from an additional unit of Ozite) will fall with the amount extracted as Ozite is used first for high-priority uses and then for lower priority ones. We assume that both benefits and costs can be measured in monetary units.

If Ozite were to be extracted for only one year, then society's interests could be expressed by saying that extraction should be such as to maximize the difference between benefits and costs. But the fact that there are multiple years involved makes matters more complicated. In this case, society's interest can be represented by maximizing the *present value* of the stream of net benefits from Ozite, where net benefits are defined as benefits minus costs and the interest rate used to discount the future represents society's trade-off between years.

This can all be described in symbols. Let

> t = the year, with the present denoted by $t = 0$;
> $x(t)$ = the amount of Ozite to be extracted in year t;
> $B(x(t))$ = the gross benefits from Ozite extracted in year t;
> $C(x(t))$ = the total cost of Ozite extracted in year t;
> r = society's discount rate; and
> V = the present value of the Ozite resource.

(For simplicity, we assume that the functions $B(x(t))$ and $C(x(t))$ do not depend directly on t and that the discount rate, r, is constant over time.) Then, the values of $x(t)$ are to be chosen to maximize:

$$V \equiv \sum_{t=0}^{\infty} \frac{\left[B(x(t)) - C(x(t)) \right]}{(1+r)^t} \qquad (1.1.1)$$

Now consider the behavior of a private, competitive market for Ozite. The mineral is being produced by a large number of small competing mines. Each mine owner will act so as to maximize the present value of its mine as seen from its own private point of view.[5] But it then follows that these actions will also lead to the socially desirable result, provided that the private and public points of view are the same.

We shall explore the proviso just given shortly. For the present, however, assume that it holds, so that we can illustrate how competitive markets operate efficiently. Suppose that there is a danger of running out of Ozite, such that a great shortage is predicted in the future if Ozite is extracted "too quickly." That expectation will surely cause the future price of Ozite to be high. But a high price for future Ozite will give private producers an incentive to reduce the rate of current extraction so as to store Ozite by leaving it in the ground.[6] This, in turn, will ameliorate the expected shortage, reduce the expected future price, and increase the present price. The process will end where the net profit to be gained from mining an additional unit of Ozite increases at the discount rate.[7]

How does this situation look from society's point of view? Evidently, the action of prices has served to ameliorate the shortage. More than that, it can be shown that V is maximized, given a finite supply of Ozite, if extraction rates are such that the additional net benefit to be gained from extracting another unit of Ozite, $[B'(x(t)) - C'(x(t))]$ (where the primes denote differentiation), grows at the discount rate, r. If all public benefits and costs are reflected in the profit-and-loss calculation of the mine owners and if society's discount rate is also the mine owners' discount rate, then the private market will reach the publicly optimal result. As with free, competitive markets generally, society's concerns are represented by the effects they have on prices, and prices serve to guide private profit seekers to a socially optimal result. Where profits are to be made, increased output is called for; where losses are envisaged, output should be reduced. Given the provisos above, free markets will do the job, and central regulation can do no better.

But of course, the provisos really matter. We can break up what is involved as follows. Private markets will lead to socially efficient results if and only if the following three conditions hold:

1. social benefits and private benefits must coincide;
2. social costs and private costs must coincide; and
3. the social discount rate and the private discount rate must coincide.

It is worth spending some time discussing at least the first two of these conditions, with particular emphasis on the matters that relate to water, especially in the Middle East.[8]

We begin with the question of the coincidence of private and social benefits. There are several ways in which this can fail, some of them more obvious than others.

First, it is possible that there are certain uses for the resource that society values beyond the willingness to pay of the users. This can certainly be true for water, where agriculture is valued beyond its private profitability or, more starkly, where there is a social interest in providing everyone, no matter how poor, with some minimal level of water for personal use.

Second, there may be certain benefits to resource use that are not captured by a private price system. In the case of water, this can happen where water is required to preserve the environment in various ways.[9] In Chile, for instance, where free water markets are used, the wetlands on which flamingoes live are drying up. A speaker on the subject at a conference asked, "Who speaks for the flamingoes?" The answer is that nobody in an unregulated private market speaks for them. Obviously, the example generalizes.

Third, as already indicated, it is crucial that the markets involved be competitive. Where there is only one or only a few large firms, private and social benefits will not coincide. The signals about consumer benefits come to producers through the prices they face. But in considering the benefits of producing an additional unit of output, a monopolist, for example, will subtract from the price the effect that additional output will have on the price of other units. Only where producers are small enough that their output decisions do not affect price will this not happen. When that is not so, private benefits will generally be smaller than social benefits.

We now turn to the question of the coincidence of social and private costs. Here the leading example involves what economists call a negative externality. Such an externality occurs when the production of a good imposes costs on others that are not brought home to the producer. Air or water pollution or other environmental effects are leading examples. In such cases, the private producer will not count such effects as costs and will produce more than is socially optimal.

There are other examples as well. In the case of water, pumping from one well can affect the costs of pumping from other wells tapping the same aquifer. Further, the costs of overpumping an aquifer will not be totally felt by any individual producer deciding how much to pump.

2. An Example: The "Rule of Capture"

Putting water aside for a moment, the following example concerning crude oil production is illuminating in more than one way.[10]

Crude oil is found in domed geologic formations in which the oil is mixed with sand. The pressure of associated gas accumulates in a cap above the oil. If a well is drilled into the side of the formation, the pressure of the gas drives the oil through the sands and out the well-hole. This is known as "primary" or "flush" production and is relatively inexpensive. Where (perhaps for the reasons described below) primary production is not feasible, resort must be had to other methods, such as secondary production, in which water is injected into the side of the formation, or tertiary production, where the oil is actually pumped. These methods are considerably more expensive.

The extraction of oil through primary production requires some care, however. If a well is drilled on the gas cap itself, then the gas will dissipate, the pressure drop, and primary production will become more difficult or infeasible. This will also happen if wells are drilled too close together. Finally, if oil is extracted at too fast a rate—above a rate termed MER, for "maximum efficient rate" of recovery—gas will exit with the oil and pressure will be lost. This phenomenon has been described by likening an oil field to an engine that will rack itself to pieces if run too fast.

Now, where the land above an oil field is divided into plots owned by different people, the question naturally arises as to who owns the oil under the surface. In the United States, this issue was settled near the beginning of the 20th century by a ruling of a Pennsylvania court. That court (which followed an earlier ruling by an English court on the ownership of percolating waters) held that oil, like percolating waters and wild beasts, was subject to the Rule of Capture. In other words, the oil belonged to whoever got it to the surface.

It is easy to see that such a ruling was disastrous because it introduced a fatal negative externality into oil recovery. The owner of the property above the gas cap had every incentive to drill without regard for the damage that such drilling would do to the cost of recovery elsewhere in the field. Decisions about well spacing were similarly made individually, without regard for the general good. Worst of all, every property owner was given an incentive to extract oil as quickly as possible. The result was that some large United States fields were ruined for flush production as property owners rushed to extract oil even though this would greatly increase everyone's costs.

Eventually, these problems were solved by a system of regulations that brought with them their own inefficiencies. For our purposes, it is more interesting to observe that they could have been solved in a different way—through a method known as unification.

Unification would work as follows. The property owners in a given oil field would form a corporation[11] that would be given control of the oil in the field. The property owners would hold stock in the corporation. Since the corporation would manage the field as a unit, it would do so taking into account all the interests involved. Since there are many oil fields in the United States, the resulting corporations would compete with each other in a competitive free market. The negative externalities would be internalized (to use the technical term), and the market solution to resource allocation would be restored.

The reader will notice, however, that this solution begs the original question of who owns the oil. Ownership rights in oil would be transformed into ownership of stock in the corporation operating the field, but how would one decide how much stock each property owner should receive?

The crucial fact here is that the answer to this question, however important to the property owners, has nothing to do with the efficient operation of the field and the market system. The operation of each field and the efficiencies gained will be the same regardless of how ownership is distributed![12] As already discussed for water, the two questions of the *ownership* of the resource and the efficient *management* or *use* of the resource, while both important, are totally independent.

Note also that unification takes account of the externality and then restores the workings of the market to achieve an efficient solution. In quite a different way, that will also be true of the methods we propose for water management.

We now return to the analysis of water.

3. Actual and Simulated Water Markets

It should now be clear that actual water markets will not lead to optimal water allocation. The reasons have already been indicated and can be summarized as follows:

- The proposition that free markets lead to an efficient allocation assumes that markets are competitive—that is, that they include a large number of independent small sellers and a similarly large number of independent small buyers. This is not typically true of water, at least in arid or semiarid countries, where water sources are relatively few and are likely to be owned by the state.
- For a free market to lead to an efficient allocation, social costs must coincide with private costs. Water production, however, involves externali-

ties. In particular, extraction of water in one place reduces the amount available in another. Further, aquifer pumping in one location can affect the cost of pumping elsewhere. Such externalities do not typically enter the private calculations of individual producers.

- Perhaps most important of all, if a free market is to lead to a desirable allocation, social benefits must coincide with private ones. If not, then (as in the case of cost externalities) the pursuit of private ends will not lead to socially optimal results. In the case of water, many countries reveal by their policies that they regard water for certain uses (often agriculture) as having a public value that exceeds its private one.

The fact that private water markets cannot be expected to lead to socially optimal results does not mean, however, that economic analysis has no role to play in the management of water systems and the design of water agreements. In particular, it is possible to build a model of the water economy of a country or region and to use that model to guide water policy. Such a model explicitly optimizes the benefits to be obtained from water, taking into account the three points made above. Its solution, in effect, provides a simulated market answer in which the optimal nature of markets is restored and serves as a guide to policymakers.

We emphasize the word *guide*. Such a model does not itself make water policy. Rather, it enables the user to express his or her priorities and then shows how to implement those priorities in an optimal way. Although such a model can be used to examine the costs and benefits of different policies, it is not a substitute for but an aid to the policymaker.

Related to this is the following point: even though the models described have their foundation in economic theory, it would be a mistake to suppose that they take only economic considerations (narrowly conceived) into account. In fact, social values and policies are of great importance in the use of such models.

We now describe the theory behind such models. We then consider how they can be used to guide decisions about water policy and infrastructure within a single country or political entity. The question of using them for the resolution of international conflicts over water will be discussed later.

4. Net Benefits from Water, Private and Social

To understand what is meant by an "optimizing model," some description of the underlying economic theory is required. We begin by temporarily ignoring some of the issues raised above and assuming for the moment that there are no social benefits from water beyond private ones. That assumption will be dropped quite soon.

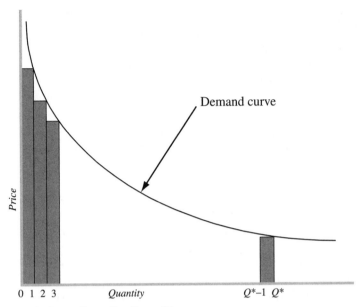

Figure 1.4.1. Gross Benefits from Water

Figure 1.4.1 shows an individual household's demand curve for water, the amounts of water (on the horizontal axis) that the household will buy at various prices (on the vertical axis). The curve slopes down, representing the fact that the first few units of water are very valuable, but later units will be used for purposes less essential than drinking and cooking.

Now consider how much it will be worth to the household in question to have a quantity of water, Q^*, as pictured in the diagram. Begin by asking how much the household would be willing to pay for the first small unit of water. The price that would be paid is given by a point on the curve above the interval on the horizontal axis from 0 to 1. (Exactly where does not matter.) So the amount that would be paid is (approximately) the area of the leftmost vertical strip in Figure 1.4.1 (one unit of water times the price in question). Similarly, the amount that would be paid for a second unit can be approximated by the area of the second-to-left vertical strip, and so on until we reach Q^*. It is easy to see that if we make the size of the units of water smaller and smaller, then the total amount that the household would be willing to pay to get Q^* approaches the area under the demand curve to the left of Q^*.[13]

Now reinterpret Figure 1.4.1 to represent not the demand curve of an individual household but the aggregate demand curve of all households in a given district. The gross (private) benefits from the water flow Q^* can thus be represented as the total area under the demand curve to the left of Q^*. These benefits are gross, however. To derive the net benefits from Q^*, we must subtract the costs of providing Q^*.

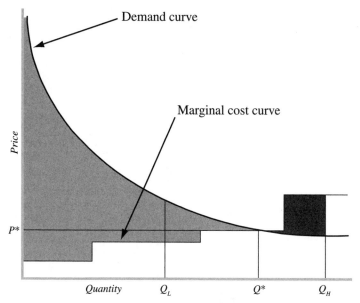

Figure 1.4.2. Net Benefits from Water

In Figure 1.4.2, the step function labeled "marginal cost" shows the cost of providing an additional unit of water. That cost increases as more expensive sources of water are used. The area under the marginal cost curve to the left of Q^* is the total cost of providing the flow, Q^*, to the households involved. Thus the net benefit from providing Q^* to these households is the shaded area in the diagram, the area between the demand curve and the marginal cost curve.

The amount of water that should be delivered to maximize the net benefits from water is Q^*, where the two curves intersect. If one were to deliver an amount Q_L, less than Q^*, then one would have a smaller shaded area, reflecting the fact that households consuming Q_L would be willing to pay more for additional units (marginal value) than the cost of such additional units (marginal cost). If one were to deliver an amount of water, Q_H, greater than Q^*, then one would have a negative value (the darker area) to subtract from the shaded area, reflecting the fact that households consuming Q_H would not be willing to pay the costs of providing the last few units. Hence, Q^* is the optimal amount of water to deliver.

As the following example shows, this apparatus can accommodate the fact that the social value of water can exceed its private value. Consider a national policy to subsidize water for agriculture by 10 cents per cubic meter at all quantities—an unrealistic but simple case. This is a statement that water to agriculture is worth 10 cents per cubic meter more to society than farmers are willing to pay for it. This is represented in Figure 1.4.3. The lower demand curve represents the private value of water to agricul-

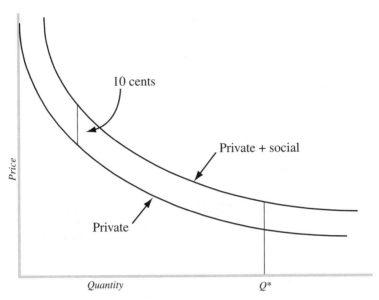

Figure 1.4.3. Social Value of Water as Revealed by a Subsidy

ture; the upper demand curve also includes the additional public value as reflected in the policy, an additional value of 10 cents per cubic meter. As this illustrates, any consistent water policy can be represented as a change in the demand curve for water. Once such a policy has been included in the demand curves, the methods used above can be used to measure net benefits.

Equivalently, optimization can take place subject to the constraints and policies imposed by the model user. In the above example, this would mean requiring that farmers must be charged 10 cents per cubic meter less than everyone else.

5. Shadow Values and Scarcity Rents

In competitive markets, prices measure both what buyers are just willing to spend for additional units of the good in question (marginal value) and the cost of producing such additional units (marginal cost). A price higher than marginal cost signals that an additional unit is worth producing, since the value buyers place on that unit is greater than the cost of production; similarly, a price less than marginal cost is a signal to cut back on production. Prices and the profits and losses they generate serve as guides to efficient (optimal) resource allocation.

As already discussed, purely private markets and the prices they generate cannot be expected to serve such functions in the case of water. Nevertheless, prices in an optimizing model play an important role—a role very similar to that which they play in a system of competitive markets.

As explained above, our Water Allocation System model, called WAS, allocates water so as to maximize the net benefit obtained from it. This maximization of net benefits is done subject to constraints. For example, at each location, the amount of water consumed cannot exceed the amount produced there plus net imports into that location.

Now, it is a general (and important) theorem that when maximization involves one or more constraints, there is a system of prices involved in the solution. These prices, called shadow values,[14] are associated with the constraints. Each shadow value shows the rate at which the quantity being maximized (here, net benefits from water) would increase if the associated constraint were relaxed by one unit. In effect, the shadow value is the amount the maximizer should be just willing to pay (in terms of the quantity being maximized) to obtain a unit relaxation of the associated constraint. (This is a topic of major importance for our analysis; a more detailed discussion with proofs and an illustration is given in the appendix to this chapter. Those unfamiliar with the shadow value concept are urged to read that appendix.)

In the WAS model, the shadow value associated with a particular constraint shows the extent by which the net benefits from water would increase if that constraint were loosened by one unit. For example, where a pipeline is limited in capacity, the associated shadow value shows the amount by which benefits would increase per unit of pipeline capacity if that capacity were slightly increased. This is the amount that those benefiting would be just willing to pay for more capacity.

The central shadow values in the WAS model, however, are those of water itself. The shadow value of water at a given location corresponds to the constraint that the quantity of water consumed in that location cannot exceed the quantity produced there plus the quantity imported less the quantity exported. That shadow value is thus the amount by which the benefits to water users (in the system as a whole) would increase were there an additional cubic meter per year available free *at that location*. It is also the price that the buyers at that location who value additional water the most would be just willing to pay to obtain an additional cubic meter per year, given the optimal water flows of the model solution. (In Figure 1.4.2, the price, P^*, would be the shadow value if Q^* were the maximum amount of water available.)

Experience shows that the following points about shadow values cannot be overemphasized:

- Shadow values are not necessarily the prices that water consumers are charged. That would be true in a purely private, free-market system. But in the WAS model, as in reality, the prices charged to some or all consumers can (and often will) be a matter of social or national policy. When such policy-driven prices are charged, the shadow values of water will reflect the net benefits of additional water given the policies adopted.

• Related to this is the fact that shadow values are outputs of the model solution, not inputs specified a priori. They depend on the policies and values put in by the user of the model.

It is important to note that the shadow value of water in a given location does not generally equal the direct cost of providing it there. Consider a limited water source whose pumping costs are zero. If demand for water from that source is sufficiently high, the shadow value of that water will not be zero; benefits to water users would be increased if the capacity of the source were greater. Equivalently, buyers will be willing to pay a nonzero price for water in short supply, even though its direct costs are zero.

A proper view of costs accommodates this phenomenon. When demand at the source exceeds capacity, it is not costless to provide a particular user with an additional unit of water. That water can be provided only by depriving some other user of the benefits of the water; that loss of benefits represents an opportunity cost. In other words, scarce resources have positive values and positive prices even if their direct cost of production is zero. Such a positive value—the shadow value of the water in situ—is called a scarcity rent.

Where direct costs are zero, the shadow value of the resource involved consists entirely of scarcity rent. More generally, the scarcity rent of water at a particular location equals the shadow value at that location less the direct marginal cost of providing the water there.[15] Just as in a competitive market, a positive scarcity rent is a signal that more water from that source would be beneficial were it available.

Water shadow values and, accordingly, water scarcity rents depend upon the infrastructure assumed to be in place.

When water is efficiently allocated, as in the solution of the WAS model, the following relationships must hold. Equivalently, if they do not hold, then water is not being efficiently allocated. (All values are per unit of water.)

• The shadow value of water used in any location equals the direct marginal cost plus the scarcity rent. For water in situ, the shadow value is the scarcity rent.

• Water will be produced at a given location only if the shadow value of water at that location exceeds the marginal cost of production. Equivalently, water will only be produced from sources whose scarcity rents are nonnegative.

• If water can be transported from location *a* to location *b*, then the shadow value of water at *b* can never exceed the shadow value at *a* by more than the cost of such transportation. Water will actually be transported from *a*

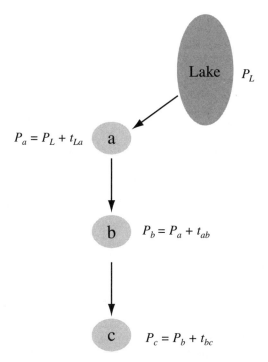

Figure 1.5.1. Efficient Water Allocation and Shadow Values

to b only if the shadow value at b exactly equals the shadow value at a plus the transportation cost. Equivalently, if water is transported from a to b, then the scarcity rent of that water will be the same in both locations.

This situation is illustrated in Figure 1.5.1, where water in a lake (L) is conveyed to locations a, b, and c. It is assumed that the only direct costs are conveyance costs. The marginal conveyance cost from the lake to a is denoted t_{La}; similarly, the marginal conveyance cost from a to b is denoted t_{ab}; and that from b to c is denoted t_{bc}. The shadow values at the four locations are denoted P_L, P_a, P_b, and P_c, respectively.

To see that the equations in Figure 1.5.1 must hold, begin by assuming that $P_a > P_L + t_{La}$ and that there is extra conveyance capacity from L to a at the optimal solution. Then transferring one more cubic meter of water from L to a would have three effects. First, since there would be one cubic meter less at L, net benefits would decline by P_L, the shadow value of water at L. (That is what shadow values measure.) Second, since conveyance costs of t_{La} would be incurred, there would be a further decline in net benefits of that amount. Finally, however, an additional cubic meter at a would produce an increase in net benefits of P_a, the shadow value of water at a. Since, by assumption, $P_a > P_L + t_{La}$, the proposed transfer would increase net benefits; hence, we cannot be at an optimum.

Similarly, assume that $P_a < P_L + t_{La}$. Then too much water has been transferred from L to a, and transferring one less cubic meter would increase net benefits. Hence, again, we cannot be at an optimum.

It follows that at an optimum, $P_a = P_L + t_{La}$, and a similar demonstration holds for conveyance between any two points.

Now, the first part of the demonstration just given requires the assumption that conveyance capacity is adequate to carry an additional cubic meter of water from L to a. Even were this not true, however, it would remain true that, in a generalized sense, $P_a = P_L + t_{La}$ at an optimum. Suppose that with the conveyance system operating at capacity, it would increase net benefits if an additional cubic meter of water could be transferred from L to a. In this case, the capacity of the conveyance system would itself have a positive shadow value measuring the additional benefit that would occur if that capacity were increased by one cubic meter. If one includes that shadow value in t_{La} (adding it to the operating costs), then the relation, $P_a = P_L + t_{La}$ is restored.

Note that shadow values play a guiding role in the same way that actual market prices do in competitive markets. As the above proof illustrates (see also the second bullet point, above), an activity that is profitable at the margin when evaluated at shadow values is one that should be increased. An activity that loses money at the margin when so evaluated is one that should be decreased. In the optimal solution, any activity that is used has such shadow marginal profits zero, and indeed, shadow profits are maximized at the optimum.

That shadow values generalize the role of market prices can also be seen from the following:

Where only private values are involved, at each location, the shadow value of water is the price at which buyers of water would be just willing to buy and sellers of water just willing to sell an additional unit of water.

Of course, where social values do not coincide with private ones, this need not hold. In particular, the shadow value of water at a given location is the price at which the user of the model would just be willing to buy or sell an additional unit of water there. That payment is calculated in terms of net benefits measured according to the user's own standards and values.

This immediately implies how the water in question should be valued. Water in situ should be valued at its scarcity rent. That value is the price at which additional water is valued at any location at which it is used, less the direct costs involved in conveying it there.

Note that the propositions about profitable and unprofitable activities involve water's being so valued. Those propositions take full account of the fact that using or processing water in one activity can reduce the amount of water available for other activities. The shadow values accompanying the optimal solution include such opportunity costs, taking into account systemwide effects. This will be particularly important in our discussion of cost–benefit analysis using the WAS model.

One should not be confused by the use of marginal valuation (the value of an additional unit of water). The fact that people would be willing to pay much more for the amount of water necessary for human life is important. It is taken into account in our optimizing model by assigning correspondingly large benefits to the first relatively small quantities of water allocated. But the fact that the benefits derived from the first units are greater than the marginal value does not distinguish water from any other economic good. It merely reflects the fact that water would be (even) more valuable if it were scarcer.

It is the scarcity of water and not merely its importance for existence that gives it its value. Where water is not scarce, it is not valuable.

6. A Word on Cost–Benefit Analysis

Plainly, shadow values and scarcity rents have a number of uses in our analysis. We point out one important use to which we shall later return.

Suppose that the conveyance line shown in Figure 1.5.1 between a and b did not exist. Suppose that when the model is run without such a line, the relation between the shadow values at a and b is such that $P_b < P_a + t_{ab}$, where t_{ab} denotes the per-cubic-meter conveyance operating cost that the nonexistent conveyance line *would* have were it in existence. (Note that capital costs are not included here.) Then such a line cannot be worth constructing under the conditions of the model run, for were such a line built, it would not be efficient to use it.

On the other hand, if the model run without the conveyance line leads to $P_b > P_a + t_{ab}$, then it would be efficient to use the line if it existed. In that case, it is worth going on to investigate whether the additional benefits brought by the line would justify the capital cost of its construction.

Evidently, then, the WAS model provides a tool for cost–benefit analysis, and we shall have much more to say about this later on. For the present, note that the model evaluates benefits taking account of the systemwide effects of a proposed project. In particular, the opportunity cost effects of a project in terms of diverted water are explicitly included in the calculation of additional benefits and, especially, in the results as to shadow values.

Appendix: Constrained Optimization and Shadow Values

In this appendix, we set out to prove the basic theorems concerning constrained optimization and shadow values. We also illustrate some of the results with a very simple nonwater example. Although we state the theorems involved in their general form, with the exception of the final theorem, we give the proofs explicitly only for the case of a single constraint. To give the general proofs merely requires a more complicated notation without adding any illumination. (The proofs remain the same but involve matrices instead of scalars.)

We begin by considering the following problem. Let x be a vector of n variables, with its components denoted by x_j, $j = 1, ..., n$. We are to choose x so as to maximize $Y = F(x)$, subject to r restrictions,

$$G^k(x) = 0, \; k = 1,...,r \tag{1.A.1}$$

It is assumed that both $F(\cdot)$ and $G^k(\cdot)$ are continuously once differentiable. It is further assumed that, at any critical point of this problem (a point at which the first-order conditions for a maximum hold), there exists an r-subvector of x such that the Jacobian of $G^k(x)$ with respect to the elements of that subvector is not zero. This assumption is maintained throughout our discussion.

Now define

$$L(x,\lambda) \equiv F(x) + \sum_{k=1}^{r} \lambda_k G^k(x) \tag{1.A.2}$$

where the λ_k, called LaGrange multipliers, are scalars whose meaning is derived below. Such an expression is called a LaGrangian.

THEOREM 1.A.1. The critical points of the original problem are identical to those values of x found by setting all the first partial derivatives of $L(x,\lambda)$ equal to zero.

PROOF. As already indicated, we give the proof explicitly only for the case of $r = 1$. (We drop the superscript on the G^k and the subscripts on the λ_k.). The condition on the Jacobian means that, in the neighborhood of any critical point, there is some element of x such that the derivative of $G(x)$ with respect to it is nonzero. Without loss of generality, assume that element is the last, x_n. Then the constraint (1.A.1) can be solved for x_n as a continuously differentiable function of the remaining elements of x. Moreover,

$$\frac{\partial x_n}{\partial x_i} = -\frac{G_j(x)}{G_n(x)}, \quad j = 1, ..., n-1 \tag{1.A.3}$$

where the subscripts denote differentiation.

Remembering that x_n is a function of the remaining elements of x, the first-order conditions for the original problem (in addition to (1.A.1)) can be expressed by differentiating $F(x)$ totally with respect to those elements and setting the result equal to zero. This yields

$$0 = F_j(x) + F_n(x)\left(\frac{\partial x_j}{\partial x_n}\right) = F_j(x) - F_n(x)\left(\frac{G_j(x)}{G_n(x)}\right), \quad j = 1, \ldots, n-1 \quad (1.A.4)$$

Now take the derivatives of $L(x,\lambda)$ and set them equal to zero, obtaining

$$0 = \frac{\partial L(x,\lambda)}{\partial \lambda} = G(x) \tag{1.A.5}$$

and

$$0 = \frac{\partial L(x,\lambda)}{\partial x_j} = F_j(x) + \lambda G_j(x), \quad j = 1, \ldots, n-1 \tag{1.A.6}$$

But (1.A.5) is just the (single) constraint (1.A.1), while (1.A.6) is the same as (1.A.4), with

$$\lambda = -\frac{F_n(x)}{G_n(x)} \tag{1.A.7}$$

This proves the theorem.

Now consider a generalization of the original problem. Suppose that some or all of the functions $F(\cdot)$ and $G^k(\cdot)$ ($k = 1, \ldots, r$) contain some outside parameter, α, so that we now write them as $F(x,\alpha)$ and $G^k(x,\alpha)$, and also write $L(x,\lambda,\alpha)$. (Of course, α need not actually be an argument of all these functions; further, the results below obviously apply when α is a vector of parameters.) We shall exemplify this below. Once x has been chosen to solve the original problem, x becomes a function of α, and so does the maximized value of $F(x,\alpha)$, which we denote by Y^*.

We now ask the question of how Y^* varies with α. One might think that this is complicated. After all, not only can α have a direct effect on $Y^* = F(x,\alpha)$ with x held constant, but also a change in α could produce a change in x. In fact, this turns out not to be the case. A highly useful result is

THEOREM 1.A.2 (ENVELOPE THEOREM):

$$\frac{dY^*}{d\alpha} = \frac{\partial L(x,\lambda,\alpha)}{\partial \alpha}$$

PROOF. Along an optimal surface, x must vary with α to satisfy the constraint (1.A.1). Inspection of (1.A.2) shows that, so long as this is true, $Y^* = F(x,\alpha) = L(x,\lambda,\alpha)$. Differentiate $L(x,\lambda,\alpha)$ totally with respect to α, to get

$$\frac{dY^*}{d\alpha} = \frac{\partial L(x,\lambda,\alpha)}{\partial \alpha} + \sum_{j=1}^{n}\left[\left(\frac{\partial L(x,\lambda,\alpha)}{\partial x_j}\right)\left(\frac{\partial x_j}{\partial \alpha}\right)\right] + \left[\left(\frac{\partial L(x,\lambda,\alpha)}{\partial \lambda}\right)\left(\frac{\partial \lambda}{\partial \alpha}\right)\right]$$

$$(1.A.8)$$

But every term except the first is zero at an optimum by Theorem 1.A.1. Q.E.D.

This result is not so surprising as it may first appear. After all, for every value of α, the elements of x are already being chosen to maximize $F(x,\alpha)$ subject to equation 1.A.1. Hence, at any such value, there is nothing to be gained (in terms of Y^*) by rearranging them. Thus, when α moves by only a very small amount, the only effect on Y^* is the direct one.

Theorem 1.A.2 is called the Envelope Theorem for the following reason. At every fixed value of α, $F(x,\alpha)$ is a function only of x. Since x is chosen to maximize $F(x,\alpha)$, $Y^*(\cdot)$ is the upper envelope of such functions.

We now consider the interpretation of the LaGrange multipliers, the λ_k. This will lead us to an understanding of shadow values.

Rewrite the constraints (1.A.1) as

$$G^k(x) = \alpha_k, \; k = 1,\ldots,r \tag{1.A.9}$$

Then, in the original problem, all the $\alpha_k = 0$. We shall adopt the convention that an increase in any of the α_k implies a loosening of the corresponding constraint, in the sense that it would allow an increase (or at least not produce a decrease) in Y^*, the magnitude being maximized. (If necessary, this can be accomplished by multiplying the corresponding $G^k(x)$ by negative one.) We can now rewrite the LaGrangian (1.A.2) as

$$L(x,\lambda,\alpha) \equiv F(x) + \sum_{k=1}^{r} \lambda_k \left(\alpha_k - G^k(x)\right) \tag{1.A.10}$$

Making the appropriate changes in notation, Theorem 1.A.2 immediately implies

COROLLARY 1.A.1:

$$\lambda_k = \frac{\partial Y^*}{\partial \alpha_k}, \quad k = 1,\ldots,r$$

Hence the λ_k are seen to be the rate at which the function being maximized increases as the corresponding constraints are loosened—the shadow values associated with those constraints.

We shall generalize this proposition below. For the present, however, it may help to illustrate the results so far obtained by means of a very simple example.

A (not very sophisticated) farmer has a piece of fencing that he wishes to arrange so as to enclose the maximum area. He is not very sophisticated because the only shape he knows about is a rectangle. In a notation restricted to this example, let the length of the rectangle be denoted by x

and the width by y. Let the entire perimeter of the rectangle (the amount of fencing) be denoted by P. Then the farmer seeks to maximize $A \equiv xy$, subject to the constraint that

$$2x + 2y = P \qquad\qquad\qquad (1.A.11)$$

Obviously, the solution is to use a square, so that $x = y = P/4$; the example is not being used to obtain this but rather because the solution is obvious.

Now consider solving the farmer's problem by the methods discussed above. Form the LaGrangian

$$L(x,y,\lambda,P) \equiv xy - \lambda(2x + 2y - P) \qquad\qquad (1.A.12)$$

Setting the derivatives of this with respect to x, y, and λ equal to zero, we obtain (1.A.11) and

$$y = 2\lambda = x \qquad\qquad\qquad (1.A.13)$$

Note that the solution of using a square is obtained. In view of this, maximized area, denoted by A^* can be written as

$$A^* = P^2/16. \qquad\qquad\qquad (1.A.14)$$

But then

$$\frac{dA^*}{dP} = \frac{P}{8} = \lambda \qquad\qquad\qquad (1.A.15)$$

where the last equality follows from (1.A.13). This exemplifies Corollary 1.A.1 and shows that λ is the shadow value of the fencing constraint. This concludes the example.

For the final set of results in this appendix, we generalize the problems so far treated and consider inequality constraints. For this purpose, we replace (1.A.1) by

$$G^k(x) \geq 0, \quad k = 1, \ldots, r \qquad\qquad (1.A.16)$$

We prove the following theorem, which, as discussed following the proof, shows clearly the existence of shadow values:

THEOREM 1.A.3 (KUHN–TUCKER THEOREM): When $F(x)$ is maximized subject to (1.A.16), there exists a set of scalars, $\lambda_k \geq 0$ $(k = 1, \ldots, r)$, such that

$$\lambda_k G^k(x^*) = 0, \quad k = 1, \ldots, r \qquad\qquad (1.A.17)$$

where x^* denotes the maximizing value of x. Each λ_k is the rate at which Y^* = $F(x^*)$ increases as the kth constraint is loosened.[16]

PROOF: Suppose that the maximum problem has been solved. At that solution, there will, in general, be some of the constraints that are binding and some that are not. Without loss of generality, renumber the constraints so that the first t are not binding. In other words,

$$G^k(x^*) > 0, \quad k = 1, ..., t$$
$$G^k(x^*) = 0, \quad k = t+1, ..., r \tag{1.A.18}$$

Let $\alpha_k \geq 0$ be defined by

$$G^k(x^*) = \alpha_k, \quad k = 1, ..., r \tag{1.A.19}$$

Consider the problem of maximizing $F(x)$ subject to the equality constraints

$$G^k(x) = \alpha_k, \quad k = 1, ..., r \tag{1.A.20}$$

Obviously, this has the same solution as the original problem. Form the LaGrangian:

$$L(x, \lambda, \alpha) \equiv F(x) + \sum_{k=1}^{r} \lambda_k \left(\alpha_k - G^k(x) \right) \tag{1.A.21}$$

From the Envelope Theorem (Theorem 1.A.2),

$$\lambda_k = \frac{\partial Y^*}{\partial \alpha_k}, \quad k = 1, ..., r \tag{1.A.22}$$

Now, for $k = 1, ..., t$, the original inequality constraints are not binding in the solution. Hence, as it were, α_k has been chosen optimally and $\partial Y^*/\partial \alpha_k$ = 0. For $k = t+1, ..., r$, on the other hand, an increase in α_k represents a weakening of the kth constraint. Since that constraint was binding, this implies $\partial Y^*/\partial \alpha_k \geq 0$, and the theorem is proved.

The conditions (1.A.17) are called complementary slackness conditions. They show that, for all k, at least one of $G^k(x^*)$ and λ_k must be zero. In particular,

$$G^k(x^*) > 0 \text{ implies } \lambda_k = 0 \tag{1.A.23}$$

which states that nonbinding constraints have their associated shadow values equal to zero. Given the nature of shadow values, that is what we should expect.

The Water Allocation System Model: A Management Tool

We now describe the Water Allocation System model in some detail and discuss some issues of special interest.

The model presented is the WAS model as used in Phase 1 of our project. It is an annual steady-state model, with extraction from water sources limited to annual, renewable amounts (which can, however, be altered by the user). Neither seasonal variation nor multiyear issues are modeled, although some of these can still be handled.

The mathematics of the annual, steady-state model (denoted WAS 3.3) are given in the appendix to this chapter, but the description of the model given here, together with Chapter 1, should suffice to provide a full understanding of what is involved. In a real sense, the true mathematical version is written in computer code.

As will be evident when it is presented, the direct treatment of agricultural demand for water in WAS 3.3 is far too simple. In Chapter 3, we discuss a much more sophisticated treatment of crop choice and water demand, the agricultural submodel (AGSM), which is intended for use with or as part of WAS, although it can also be (and has been) used independently.

For convenience, we refer to the area whose water economy is being modeled as a country. In fact, WAS modeling can apply to regions larger or smaller than an actual country.

Most important of all, even for a given country, WAS is not a single model fixed in stone. Rather, it is a tool that the user can employ to explore the consequences of various decisions and alternate circumstances.

1. Private Water Demand

The country is divided into districts. Typically, districts are chosen in terms of data availability, but in principle, they should be as small as possible, since it will be assumed that intradistrict conveyance costs per cubic meter are constant (and small). Although it is not technically necessary that the districts defined for demand purposes be the same as those used for supply, we shall assume that they do coincide.

Within each district, private water demand is divided by user type. In the model as used in the Middle East, the user types are taken to be households, industry, and agriculture (and separate provisions are made for environmental uses), but other divisions are permitted.

The model will be used to investigate different demand conditions, and it is natural (but not inevitable) to refer to these as the conditions that existed in some base year (in our case, in 1995) and that are forecast for future years (in our case, for 2010 and 2020). Obviously, such forecasts can vary, and WAS permits the user to take this into account.

As already indicated in Chapter 1, specification of demand here does not merely mean specifying the quantity that will be demanded. Rather, it means specifying the demand *curve* that shows water demand as a function of price. This is particularly important for agriculture and industry but applicable to all user types. In the present version of WAS, such specification is done as follows (although, as already remarked, agriculture receives a special treatment).

Let Q denote the quantity of water demanded by a given user type in a given district in cubic meters. Let P denote the price per cubic meter charged to such users. Then the demand curve of those users in the relevant range is approximated by

$$Q = AP^\eta \qquad\qquad (2.1.1)$$

where $A > 0$ and $\eta < 0$ are parameters whose value will be as described below. Such a demand curve has a constant price elasticity, where price elasticity is defined as $(d \log Q/ d \log P)$, or approximately, the percentage change in quantity induced by a 1% change in price. For expository convenience, demand price elasticity is measured positively, so that in equation 2.1.1, it is given by the value of $-\eta$.

Obviously, such a curve can be only an approximation in some range, particularly if η is close to zero. Otherwise equation 2.1.1 would imply that users would pay an unbounded amount of money for vanishingly little water as the price increased.[1] Hence, care must be taken in interpreting the results of WAS runs if prices are very high. However, for prices in the general range of historical experience, equation 2.1.1 appears a reasonable approximation for most users—particularly in view of the sensitivity analysis described below. The general WAS methodology (as opposed to its current implementation) does not require the use of the constant-elasticity form.

Although the values of η can be altered by the user of the model, in the Middle East implementation of WAS, the default values of η are set (with some reference to the scanty literature) at fairly low magnitudes: 0.2 for households, 0.3 for industry, and 0.5 for agriculture. Since the price elasticity of water demand is certainly low in the relevant range, the results of WAS runs are not very sensitive to the precise value of η chosen.

The important parameter in terms of results is A, which defines the position of the demand curve rather than its (logarithmic) slope. We now consider how A is to be specified.

Choose a particular price for water, say P_0. Consider the quantity of water, Q_0, that would be demanded by the given user type in the given district if users were charged P_0 and *if there were no restrictions on the amount of water available.* In the case of a base year already past, this will often mean choosing P_0 as the price actually charged and Q_0 as the amount actually consumed. For future years, the matter will generally require more thought.

Once η, P_0, and Q_0 have been specified, the value of A can be calculated from equation 2.1.1.

One must beware, however. The italicized proviso is extremely important; experience has shown that there is a tendency to take Q_0 as the amount consumed in the base year even if that amount was determined by restrictions on water supply. One must not confuse supply and demand. The running of WAS will determine whether quantities demanded can actually be supplied. Where supply restrictions constrain water consumption, then, even in the base year, further analysis is required.

Such analysis need not be very complicated. Begin with the case of households for some future year. Here it is obviously sensible to break the problem into two parts: forecasting the size of the population and forecasting per capita (or perhaps per household) demand at P_0. Population forecasts are usually available from outside sources, and they are what will really influence the results. Since the value of A can be altered by the user, it is easy enough to examine the results of varying the population size. Given per capita demand, population size is proportional to A.

Per capita demand can be estimated by considering forecast changes in household income and applying an adjustment for increasing income to base-year consumption, if that consumption was unrestricted. If base-year consumption was restricted, then using the per capita consumption of other countries that did not have such restrictions and adjusting for income will produce an acceptable estimate. (Remember that fine-tuning per capita demand is unlikely to have much effect on the results.)

Industrial water demand is somewhat harder to estimate but, at least for the Middle East, also less important. Basically, this requires forecasting the growth of industry and the extent to which water would be demanded for production processes at a particular price if water supply were unrestricted.

Agricultural demand can be similarly handled, but here there is more need for care. This is true for several reasons:

- Agriculture is usually the largest water-using sector.
- There is likely to be considerable public interest in policies that ensure water supply to agriculture, so accurate representation of demand may be particularly important here.
- The price elasticity of demand for water seems likely to be higher for agriculture than for the other two user types, making the results sensitive to agricultural demand specification.
- Finally, agriculture can typically use more varieties of water quality than can be used by industry or especially households.

As already mentioned, we give special treatment to agriculture in the AGSM model discussed in Chapter 3, but we must now discuss the last point made above as to different water qualities, particularly since WAS 3.3 does not make direct use of AGSM.

The specification and discussion of the demand curve equation 2.1.1 effectively assumes that there is only one quality of water or that all water types are perfectly interchangeable. This is certainly not true. How then should demand relationships be specified? There are several cases to consider.

Begin with the case of households. If, as is usually the case, households use only potable water, the existence of different water qualities does not complicate the specification of household demand. The fact that potable water can be produced in different ways from different sources (desalination of brackish water or treatment of surface water, for example) can be dealt with as a matter of supply and technology. Such production involves the different ways in which potable water can be supplied to the household; it has nothing to do with the demand of the household itself for potable water.

If households can use nonpotable water for certain uses, then two household demand functions must be specified. The first is for potable water, used for drinking and cooking, and is specified as before. The second is the demand function for household uses that do not require potable water, such as watering lawns and washing cars. If these lower priority uses involved only nonpotable water, then there would be no difficulty. The problem comes because potable water can also be so used—and will be if it is sufficiently cheap.

Here, the simplest thing to do is to specify the demand curve for lower priority uses as in equation 2.1.1 but then to assume that potable and nonpotable water are perfect substitutes in such uses. (This can be done either by considering them totally interchangeable or, as in the case discussed for agriculture below, by assuming that the household bears an implicit cost per cubic meter when one water type is used rather than the other.) Which water type gets used in fact will then depend on the prices faced by the household (and on the implicit cost, if any, of using a particular water type).

The case of industry can be handled in the same way. If certain uses require potable water, then the demand for those uses must be separately specified. If all types of water are interchangeable (perhaps with implicit costs of use), then demand for them can be treated as demand for a single good that may be sold at different prices.

As already stated, agriculture presents a more formidable problem. This is because crop yields often depend on water quality, making the perfect-substitute assumption unrealistic. In WAS 3.3, we have nevertheless used the same procedure as for lower priority household uses, noting that here there can be either a positive or a negative cost (a benefit) from the use of treated wastewater. That cost can be positive because of impurities that affect equipment or crops or negative because of nutrients in the water.

Plainly, this is not an adequate treatment, and we therefore devote considerable effort to a separate treatment of agricultural demand—one that explicitly looks at the effects of water quality on crop yields.

Note that restricting agricultural use of treated water for environmental reasons is not a matter of *private* agricultural demand for water. Such environmental concerns have to do with externalities, and we discuss them below.

Now, it may be asked (particularly by economists) why we do not simply expand the specification of equation 2.1.1 to account for different water types. Suppose, for example, that there are two types. Denoting water types by subscripts (1 and 2),[2] why not write

$$Q_1 = A_1 P_1^{-\lambda} P_2^{-\mu}$$
$$Q_2 = A_2 P_2^{-\gamma} P_2^{-\delta} \tag{2.1.2}$$

and similarly for more than two types?

The answer is that the number of the parameters and especially the relationships among them quickly becomes unmanageable if this is done. Standard results of the economic analysis of the firm or the household show that the six parameters in equation 2.1.2 cannot be independently specified. Further, the available data will typically not support their estimation by econometric methods. Hence the approximations already discussed seem preferable.

2. Naturally Occurring Supply

Water supply consists of naturally occurring water (rivers, lakes, and aquifers), treated wastewater, and desalinated seawater. We discuss the treatment of naturally occurring water first.

Naturally occurring water is modeled as follows: Within each district are a number of water sources (some or all of which it may be convenient to aggregate). It is assumed that each water source has an annual renewable amount and a constant extraction cost per cubic meter up to that amount. The annual renewable amounts and the per cubic meter extraction costs are inputs to WAS. Both require discussion, and we begin by making several points about the simpler one, extraction costs:

1. The assumption of constant per cubic meter extraction costs is a convenience rather than a necessity. Permitting average costs to vary with output would present no great difficulty.
2. Any necessary treatment costs can simply be included with extraction costs. This holds for desalination of brackish water as well as for treatment for biological problems.
3. Note, however, that the costs involved are all operating costs. The capital costs involved in the installation of treatment facilities are not included here if the facilities already exist. An extensive discussion of the handling of capital costs is given later.

We now turn to the discussion of annual renewable amounts. Here there are the following things to consider:

1. The default value of the annual renewable amount from a given source is the average annual renewable amount. But this can be varied by the user. In particular, dry or wet years can be readily examined in WAS 3.3 by multiplying all annual renewable amounts by some constant (less than 1 for a dry year, greater than 1 for a wet one).
2. On the other hand, the fact that WAS is currently an annual, steady-state model means that it is difficult to use WAS 3.3 to examine the effects or implications of changing temporal patterns of rainfall and policies of aquifer and other resource management that take such changes into account. Such investigation awaits the development of a multiyear model.
3. The annual renewable amount should be considered the amount that can be extracted given the hydrology involved. This should not be confused with the capacity of the existing extraction facility, since that capacity is a matter of choice. (See point 6 below, however.)
4. Perhaps most important of all, different sources can draw on the same water resource—for example, the same river or the same aquifer.[3] Where the sources are all in the same district and have the same per cubic meter extraction cost, this is unimportant, since they can simply be aggregated. Where, however, they have different extraction costs or are in different districts, this "common pool" phenomenon must be handled.[4] We deal with it by constraining the amount of water that can be extracted from a given source in a given district in more than one way. The first such constraint is the one already discussed: extrac-

tion must not exceed the annual renewable amount already specified. The second constraint is the common-pool constraint: total extraction from all sources drawing on the same water resource cannot exceed the annual renewable amount of that resource.

5. When such constraints are specified, the WAS optimization will decide the efficient geographic extraction pattern. Note that this is not merely a matter of relative extraction costs. The efficient extraction pattern will also depend on the geographic pattern of demands, on the availability of other water resources, and on the conveyance system assumed to exist.

6. Annual renewable amounts do not seem naturally to apply to the case of fossil aquifers. The simplest way to handle that case is to limit annual extraction by the capacity of the extraction facilities (contrary to point 3 above). The choice of that capacity can remain an infrastructure choice whose costs and benefits can be calculated as later described.

The problem is that the solution takes no account of the fact that a fossil aquifer is a nonrenewable resource, and extraction of water in one year reduces the total amount available for extraction in later years. A really satisfactory treatment requires a multiyear analysis. Unless the capacity of the extraction facilities is large relative to the size of the fossil aquifer, however, the problem can be handled as above. In our case, the most important fossil aquifer is the Disi aquifer in Jordan (see Chapter 7), where extracting at planned capacity could take place for a century, making the depletion problem negligible at any reasonable discount rate.

3. Treated Wastewater

WAS permits the inclusion of treated wastewater. This is done by specifying the existence (or proposed existence) of treatment plants and associated conveyance systems. That specification is done as discussed below in the section on infrastructure, but there are a number of additional matters of interest.

In the current model, it is assumed that while households and industry generate wastewater, only agriculture can use the treated wastewater. (That assumption can easily be changed, but we shall continue it in the following discussion.) Not all effluent produced by household and industry becomes available for treatment; the user of the model must specify what fraction of the effluent is so available.

We assume that wastewater must, in any case, be treated until it is safe for discharge into the environment, or environmental damage will result. Accordingly, a per cubic meter penalty for this is levied against the use of fresh water by households and industry. Such an effluent charge need not appear in the water prices actually charged to such consumers; prices can be a matter of policy, as described below. The effluent charge is rather an implicit penalty to be subtracted from the net benefit function maximized by the model. Nevertheless, it is instructive to examine the effect of the effluent charge on the shadow values that are part of the optimal model solution.

Let the per cubic meter effluent charge be denoted by E. Suppose that prices to an effluent-producing set of users are not fixed by policy but that the amount of water they receive is simply to be determined by WAS. Then one can read an implicit price from examination of the demand curve for that user group, solving equation 2.1.1 for P in terms of Q. This is, indeed, the price that, if charged to the users, would cause them to demand exactly the allocated quantity. Call that price P^*. Let V denote the shadow value of fresh water in the given district. Then

$$P^* = V + E \tag{2.3.1}$$

In other words, the effluent producers are treated as though they were directly assessed the effluent charge.

This is as it should be. Allocating an additional cubic meter of water to a noneffluent-producing consumer in the district costs the system V, the shadow value and opportunity cost of the water, which then cannot be used by others. Allocating it to an effluent-producing consumer means incurring the additional environmental penalty measured by E. In a private market system, this would be appropriately reflected by increasing the price paid by effluent-producing consumers by exactly E. That would be the exactly appropriate disincentive to consumption needed. In the WAS model, the same thing is indicated by equation 2.3.1.

Now, we have so far not discussed how the value of E should be set. In districts where an existing treatment plant is not used to capacity, this is an easy matter: E should be set at the per cubic meter treatment cost involved in bringing effluent to an environmentally safe standard. (Note that the capital costs of the plant are not to be included in E. See the discussion below.)

In other districts, the matter is not so simple: at least some effluent will remain untreated, causing environmental damage. It would be simplistic to argue that such damage cannot exceed the costs of treatment per cubic meter (this time including capital costs), or else treatment capacity would already have been built or expanded to treat the effluent, but there is something in this argument in the long run. In our use of WAS, we have set the effluent charge in all districts at the operating costs of treatment, even though it is unclear whether this is correct for districts with inadequate treatment capacity or those with no treatment plants at all.

Wastewater need not be treated only to make it safe for environmental disposal, however; with additional treatment, it can be made safe for use in agriculture. In WAS, these additional per cubic meter costs are assigned to the use of wastewater in agriculture. (Note that it would be inappropriate to assign E in this way; effluent costs are incurred by the production of effluent, whether or not additional treatment is performed.) It will be efficient to use such additionally treated wastewater whenever the shadow value of wastewater exceeds the additional per cubic meter treatment costs. Indeed, such use will occur in the optimal solution up to the capacity of the treatment plant.

It is interesting to see what becomes of equation 2.3.1 in this case. Let T be the operating costs of additional treatment per cubic meter and again assume that the effluent-producing user group does not have the prices charged to it specified by the user of the model. Let M denote the fraction of effluent that is available for treatment. Let V' be the shadow value of treated wastewater. Then, assuming that the capacity of the treatment plant is not reached,

$$P^* = V + E - M(V' - T) \tag{2.3.2}$$

It is easy to see why this should be so. The production of effluent now has two effects. The first of these is the environmental cost, E, as before. The second, however, can be a positive effect. If the shadow value of additionally treated wastewater exceeds the cost of additional treatment, then additionally treated wastewater has a positive scarcity rent. In this case, the production of effluent has a positive effect in that it permits more of the valuable, additionally treated wastewater to be produced. This (generally partially) offsets the environmental effect and reduces the extent to which consumption of fresh water by effluent producers should be discouraged.

4. Infrastructure: Capital Costs

WAS permits the user to specify the water-related infrastructure that is in place. Such infrastructure can already exist or be under consideration and includes several types:

- wastewater treatment plants;
- desalination plants;
- interdistrict conveyance links (for either fresh water or treated wastewater);
- treatment plants for naturally occurring water; and
- additional or expanded treatment facilities.

In all cases, the user must specify both the capacity of the infrastructure and the per cubic meter cost of using it. As we shall now see, the latter specification raises some delicate issues.

Obviously, the capital costs of the infrastructure project are important and must be dealt with in some fashion, but the particular fashion matters a good deal. In reading the following discussion, the reader should bear in mind that we are not departing from the view that efficient management means that capital costs must be recovered. The issue is rather how this should be done in the WAS analysis.

We begin by considering the case of an infrastructure project that has already been built. (In most cases, this will be an existing project, but for purposes of analyzing other projects well in the future, it might be a project not yet built that will have been completed by the construction time of other such projects.) Here the capital costs have already been expended.

In such a case, it would be a mistake in the use of WAS to attribute those sunk capital costs to the use of the infrastructure when examining efficient water allocation. This is true even if the capital costs are being recovered in the per cubic meter prices actually charged to water users.

The reason for this is as follows. The capital costs of an existing piece of infrastructure have already been expended. Those costs will be neither increased nor diminished if the infrastructure is used. It follows that, were one to assign those costs as part of the per cubic meter costs of using the infrastructure, WAS would take into account a disincentive to a use that has no basis in reality. This would serve to distort the optimization process and result in the wrong set of water flows.

An example may help here. Suppose that there is an existing desalination plant. Suppose that desalination technology has improved since the plant was built such that the per cubic meter cost of a new plant, including capital costs, would be lower than the per cubic meter cost of the existing one, again including capital costs. Suppose, on the other hand, that the per cubic meter cost of the new plant, including capital costs, would be higher than the per cubic meter operating cost of the existing one, this time *excluding* capital costs. Then (assuming the existing plant has adequate capacity), the efficient thing is to go on using the old plant and not build the new one. This is the answer that WAS will find if the sunk capital costs of the old plant are not put in as part of the cost of its use. On the other hand, if those capital costs are included, the recommendation will be to scrap the old plant and build the new one, and this is clearly wrong.

Another way to say this is to observe that including the already expended capital costs of the existing plant transforms the choice from using the existing plant or building a new one to deciding which of the two plants should be constructed, if the existing plant did not already exist. That is the wrong question, and asking it will lead to incorrect results.

The same analysis applies in the case in which one is considering the construction of a given piece of infrastructure in the presence of other projects that are assumed to have been constructed before the given one. There the question asked presumes the existence of those other projects. It is not the same as the question of whether the given project should be constructed if the other projects are not, nor is it the same as the question of which potential projects (the given one plus others) should be built. For the question posed, the projects assumed to have been built first must not have capital costs assigned to their use. Of course, the capital costs of proposed projects must be treated, and we shall do so in the following section.

Finally, consistent planning requires that the capital costs of all projects be recovered (either through charges or through subsidies), and that requirement neither ceases when the project is built nor fails to apply to existing projects. But this is a matter not of costs but of prices. The user is free to include the capital costs of existing infrastructure in the prices charged to water users; this is done making use of the policy facilities of WAS, discussed below. We observe, however, that if existing projects are not used to capacity, it may be inefficient to recover their capital costs through a per cubic meter charge that gives consumers a disincentive to their use. In such cases, it may be better to recover capital costs through a separate hookup charge that does not affect the water consumption decision directly.

One more point before proceeding. The shadow value of water that uses a particular project will include the shadow value of the capacity of that project. That shadow value will be zero if project capacity is not reached but positive if project capacity is a binding constraint. In a multiyear optimization model in which the timing of capital expenditures and project construction is handled explicitly, the present discounted sum of the shadow values of the capacity constraint would equal the capital cost of expanding capacity by one unit.[5] In such a context, much of the problem discussed in this section would not arise.

5. Infrastructure: Cost–Benefit Analysis

We now turn explicitly to the question of how WAS facilitates the cost–benefit analysis of proposed infrastructure projects. It is most important to note in this regard that the benefit calculations in WAS include all the system-wide effects of the proposed project. In particular, if the project will—as is very often the case—change the flows of water and thus benefit some users while harming others, WAS will take that into account, using the opportunity costs of the diverted water.

WAS 3.3 handles the linked issues of (1) cost–benefit calculations for proposed new infrastructure and (2) capital costs in two alternative ways: as a lump sum or on a per cubic meter basis.

Capital costs as a lump sum: Direct cost–benefit analysis

The recommended method for dealing with capital costs of proposed infrastructure is to make a direct comparison of benefits and costs. In this method the capital costs are not included in the data on a per cubic meter basis. The model is run with and without the proposed infrastructure and the difference in net benefits calculated. Because that difference is an annual one, the stream of such differences[6] must be capitalized so that its present value can be compared with the capital costs of the project.

This method has both advantages and defects. In particular:

- The method does not require the separate calculation of per cubic meter capital costs, which depend on the actual quantity used, and is thus very easy to use.
- The shadow values in the scenario with the projected infrastructure will not accurately reflect all costs involved because they will not include capital costs.
- The method will not choose which projects to build from a menu of possible projects.

Capital costs per cubic meter: Indirect cost–benefit analysis

An alternative method is to include the per cubic meter capital cost of the proposed infrastructure as a cost associated with the use of that infrastructure. If the new infrastructure's benefits exceed its costs (including capital costs) in the assumed steady-state conditions, then the model will use the infrastructure. Otherwise it will not.

Note that, as discussed above, the per cubic meter capital costs of existing infrastructure (or of infrastructure that will exist by the time the proposed project is built) must never be included in this way. However those capital costs are recovered in the charges made to users, they are sunk costs. Including them as use-related costs will lead the model not to use existing infrastructure fully because it will wrongly assume that it can save capital costs by so doing.

This method is related to a very useful property of the model: if it is run without the proposed infrastructure in question, then the shadow values obtained provide per cubic meter cost targets for the proposed infrastructure. Two examples will make this clear.

- If the model is run without desalination plants, the shadow values of water at the possible plant sites give the per cubic meter target values for desalination cost (including capital cost).
- More generally, if one is considering a new source of water, the shadow value of water obtained at the location of the proposed new source gives the per cubic meter cost target (including capital costs) that the new source will have to meet to be worth using. Note, in particular, that this is true of a proposal to import water. Here the shadow value of water at the proposed place of entry gives the per cubic meter target at which

the imported water must be available to be worth having. That target is a delivered-price target with all conveyance costs (including capital costs of conveyance infrastructure) of bringing the imported water from the point at which it is sold and its purchase price there.

As with the previous method, the alternative method has both advantages and defects. In particular:

- The per cubic meter capital costs of the new infrastructure will show up in the shadow values of water. Since such values are expressed in units of dollars per cubic meter, they bear a strong resemblance to prices. The inclusion of capital costs will make the values look relatively natural in terms of prices.

- On the other hand, the shadow values are not the same as the prices charged to users. It is a mistake to think of them in that way. Further, since the capital costs of existing infrastructure (or of infrastructure that will exist by the time the proposed project is built) must not be included (see above), the shadow values will not include them, so they will in any case not resemble the prices charged to users.

- The fact that the shadow values of water in each location will include the capital costs of the new project does, however, make those shadow values more accurate as measures of the value of the marginal cubic meter than are shadow values that do not include such costs. For example, if the proposed infrastructure is a conveyance line, the shadow value of water at the terminus should properly reflect the fact that if water were delivered there by other methods, then the capital cost of the proposed infrastructure would be saved.

- The big advantage to using the method described is that one need not test out only one proposed project or collection of proposed projects at a time. By specifying the costs of all projects under consideration and seeing which ones are used in the model solution, one can tell which of a menu of projects should be built in the presence of the others (under the assumed steady-state conditions.)

- On the other hand, this method will not readily handle the cost–benefit analysis of the capacity expansion of existing infrastructure. For example, when conveyance capacity between two points is to be expanded with an additional line, the capital cost of the new line is not incurred when water moves over the old one. Just increasing the capacity and averaging the new and old costs on a per cubic meter basis will therefore give the wrong result. Although this can be dealt with, the first method described above handles it easily.

- The method under discussion will not generate the total benefits of the new infrastructure for direct comparison with its capital costs. But it can be made to generate estimates of both the annual net benefits and the present discounted value of the net benefits, where *net* means "net of capital costs."

It should be emphasized that both methods are correct. If properly used, they are equivalent, and the choice between them becomes largely a matter of convenience.

6. Public Policies Toward Water

The first method of cost–benefit analysis—running the model with and without the proposed infrastructure—can be generalized to permit the examination of the costs and benefits of particular public policies. This can be done by running the model with and without the policy or policy change being examined and observing the change in net benefits as well as any direct costs to the government from adopting the policy.

We now discuss the policies that can be imposed or examined in WAS 3.3.

Price policies

The most important set of policies concerns the prices to be charged to water users. Here, WAS 3.3 permits three types of price policies, the first two somewhat simplistic and the third more realistic.

First, the model user can specify that one or more particular groups of users in one or more districts are subsidized and pay a specified, fixed amount per cubic meter less than would be indicated by the shadow value of the water they use.[7] If that amount is 10 cents per cubic meter, for example, then this is an explicit statement that the social value of water going to those users is 10 cents per cubic meter greater than its private value. (See the discussion in Section 4 of Chapter 1, especially Figure 1.4.3.)

Note that the subsidy can be negative. This would amount to a tax on the specified water users and would be a statement that the social value of water going to them was less than its private value.

Second, the model user can specify that one or more particular groups of users in one or more districts are subsidized and pay a specified percentage amount per cubic meter less than would be indicated by the shadow value of the water they use. This makes a percentage statement about the relation between social and private value. Again, the "subsidy" can be a tax.

Finally, the model user can specify the specific prices to be paid by particular user groups in particular districts. This can be done in terms of a step function in which a certain amount is paid per cubic meter for a first specified quantity of water, a different amount for a second specified quantity, and so forth. The price schedules specified for fresh water and for treated wastewater can be different. This amounts to a specific statement about the social value of water going to these users.

Such schedules have one problem. In reality, schedules are specified at the level of the individual water user (a single farm, for example). Because the model operates at the district level, the schedules must be specified at that level. The most convenient way to do this—by assuming that all users in the specified group will be alike in water consumption and multiplying the size of the quantity steps for an individual user by the number of users in the group—is not completely accurate. If some users are larger than others (in size of farm, for example), then this can be taken into account (using land areas, for example).

Again, however prices are specified, the WAS optimizing solution will respect those prices. One must not suppose that the shadow values of water and the prices charged to users invariably coincide.

It may be useful at this point to consider whether shadow values and user prices *should* coincide (after adjustment for effluent charges and wastewater treatment profits). If they do, then charging such prices to users will result in their demanding exactly the water flows that the model shows to be optimal. At first sight, that seems a powerful argument for setting prices in this way.

There are, however, two excellent reasons for not doing this. In the first place, as discussed in Section 4 above, the capital costs of existing infrastructure will not be reflected in the shadow values of water, while the capital costs of projected infrastructure will. But consistent planning surely requires that capital costs be recovered, and this cannot be avoided merely because the planning takes place after some of the infrastructure is in place. Hence shadow-value pricing will not recover the capital costs of existing structures, and this is unacceptable. One might think that this problem can be avoided by recovering capital costs through hookup charges or other charges that are not on a per cubic meter basis. (See Section 4 for the argument in favor of doing this.) Unless this is done for existing structures only, however, the projected prices for water using planned infrastructure will be too low and the flows too large relative to the efficient optimum.

The second reason for not using shadow-value pricing is more fundamental and has already been discussed. Certain users may attach social values to the consumption of water that are not reflected in private values. Although, as discussed below, this can be handled by specifying restrictions on the quantities of water that go to these users (minimum amounts, for example), such restrictions already depart from a purely private system (as they must) and will affect shadow values. When, by individual or collective judgment, water is found to be not purely a private good, it may be simpler to specify the prices to be charged directly.

We add one note, however. Altering water prices may not always be the most efficient way to effect social aims. If what is wanted is to maintain agriculture, for example, it may be better to subsidize agricultural *output* directly rather than interfering with the efficient *input* choice of farmers. Of course, traditional or social reasons may argue against such a policy.

Environmental policies

Environmental issues are, of course, very important in water management and policy. Although the treatment of such issues in WAS 3.3 is not as complete or sophisticated as one might wish, there are nevertheless several ways in which environmental concerns can be expressed. We first briefly discuss these and then consider how a more adequate treatment might be accomplished in the context of a multiyear version of WAS.

- WAS limits extraction from a source to a specified annual, renewable amount. Unless such amounts are set deliberately high by the user, this prevents overpumping of aquifers and hence any consequent deleterious effects on water quality.
- An effluent charge can be imposed on those uses of fresh water that create effluent.
- The model user can specify (1) whether and where treated wastewater can be used, (2) a maximum percentage or amount of wastewater that can be used in irrigation in a particular district, and (3) a per cubic meter penalty (or benefit) for the use of treated wastewater.
- The user can specify that certain amounts of water be set aside for unspecified purposes in certain districts. One of the purposes of such specification is to allow the user to reserve water for environmental purposes.

Now, it should be noticed that the various possibilities all represent the use of policy instruments to deal with environmental concerns. The environmental concerns themselves are not directly modeled. Hence the user must decide on the appropriate choice and level of the policy instrument to be used without full internal guidance from the model. Although use of WAS can reveal the systemwide costs of different levels of policy action, it cannot reveal the different benefits that result.

To some extent, this problem would be alleviated in a multiyear treatment, in which one might model the ways in which decisions in one year affect water quantities and qualities in later years. If one can also model the effects on benefits of water quality (as well as water quantity), then at least those environmental effects can be endogenized. For example, if one knows how overpumping of aquifers affects the salinity of groundwater in later years and also knows how agricultural yields are affected by salinity, then, in a multiyear model, the decision of whether or how much to overpump an aquifer can be made taking into account the effect on agriculture in later years. That calculation would be made within the optimization routine of the model itself rather than have to be made exogenously by the user. A similar statement holds for at least some of the effects of effluent discharge or of the use of treated wastewater in agriculture.

Not all environmental issues could be handled this way, however. For example, the effect of river water levels and quality on fish, animal, and plant life could not be so easily treated. This is not because such effects cannot be quantified and modeled; in principle, they can be. Rather, it is because the value to be assigned to such matters is difficult to decide and make fully commensurable with the private values measured and accounted for in the model. Failing that, one must deal with such matters directly in terms of the instruments to be used, and that is what is done in WAS 3.3.

Other policies

WAS 3.3 also permits the user to impose other policies. We have already seen that the user can specify that certain amounts of water must be reserved in certain districts. In the discussion just given, we supposed that such set-asides were for environmental purposes, but this does not have to be the case. For example, where a treaty specifies that a certain amount of water must be delivered to a neighbor at a particular place, the set-aside facility can be used to model that obligation. (The receiving country can model the same phenomenon as a separate source of supply.)

The user can also specify that certain users in certain districts must receive at least some minimum amount of water. This is readily accomplished by offering those users a low fixed price for the minimum amount. (Note that as discussed above, the minimum amount must be calculated at the district level, not the level of the individual user.)

Alternatively, it may be thought that there would be social costs (or costs not well represented in short-run private demand curves) if certain consumers received relatively little water. Instead of prescribing a minimum amount of water to be delivered to such consumers, the model user can specify a per cubic meter penalty to be recognized by the model if the delivered amount falls short of some target.

Appendix: The Mathematics of the WAS Model

The WAS model is written in the Generalized Algebraic Modeling System (GAMS) language. As the objective function is not linear, the MINOS nonlinear operating system is used to solve the WAS model. The model is presented below in the standard form for optimization, namely, the objective function followed by the constraints. In mathematical terms,[8] the model is as follows[9]:

$$\max Z = \sum_i \sum_d \left(\frac{B_{id} \times \left(QD_{id} + QFRY_{id} \right)^{\alpha_{id}+1}}{\alpha_{id}+1} \right) - \sum_i \sum_s \left(QS_{is} \times CS_{is} \right)$$

$$- \sum_i \sum_j \left(QTR_{ij} \times CTR_{ij} \right) - \sum_i \sum_j \left(QRY_{ij} \times CR_{ij} \right)$$

$$- \sum_i \sum_j \left(QTRY_{ij} \times CTRY_{ij} \right) - \sum_i \sum_j \left[CE_{id} \times \left(QD_{id} + QFRY_{id} \right) \right]$$

Subject to[10]:

$$\sum_d QD_{id} = \sum_s QS_{is} + \sum_j QTR_{ji} - \sum_j QTR_{ij} \quad \forall i$$

$$\sum_d QFRY_{id} = \sum_d QRY_{id} + \sum_j QTRY_{ji} - \sum_j QTRY_{ij} \quad \forall i$$

$$QRY_{id} = PR_{id} \times \left(QD_{id} + QFRY_{id} \right) \quad \forall i,d$$

With the following bounds:

$$\left(QD_{id} + QFRY_{id} \right) \geq \left(\frac{P_{\max}}{B_{id}} \right)^{1/\alpha_{id}} \quad \forall i,d$$

$$QS_{is} \leq QS_{\max is} \quad \forall i,s$$

$$PR_{id} \leq PR_{\max id} \quad \forall i,d$$

all variables positive, where:

Indices

i = district

d = demand type (urban, industrial, or agricultural)

s = supply source or steps

Parameters

α_{id} = Exponent of inverse demand function for demand d in district i

B_{id} = Coefficient of inverse demand curve for demand d in district i

CE_{id} = Unit environmental cost of water discharged by demand sector d in district i ($\$/m^3$)

CR_{id} = Unit recycling cost of water supplied from demand sector d in district i ($\$/m^3$)

CS_{is} = Unit cost of water supplied from supply step s in district i ($\$/m^3$)

CTR_{id} = Unit cost of water transported by demand sector d in district i ($\$/m^3$)

$CTRY_{id}$ = Unit cost of recycled water transported by demand sector d in district i ($\$/m^3$)

$P_{\text{max } id}$ = Maximum price of water from demand sector d in district i

$PR_{\text{max } id}$ = Maximum percent of water from demand sector d that can be recycled in district i

$QS_{\text{max } is}$ = Maximum amount of water from supply step s in district i (mcm)

P_{id} = Shadow value of water for demand sector d in district i (computed) ($\$/m^3$)

Variables

Z = Net benefit from water in million $\$$

QS_{is} = Quantity supplied by source s in district i in mcm

QD_{id} = Quantity demanded by sector d in district i in mcm

QTR_{ij} = Quantity of freshwater transported from district i to j in mcm

$QTRY_{ij}$ = Quantity of recycled water transported from district i to j in mcm

QRY_{id} = Quantity of water recycled from use d in district i in mcm

$QFRY_{id}$ = Quantity of recycled water supplied to use d in district i in mcm

PR_{id} = Percent of water recycled from sector d in district i in mcm

A. Freshwater continuity

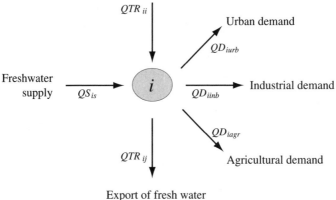

Import of fresh water
from other districts

QTR_{ii}

Urban demand

QD_{iurb}

Freshwater
supply QS_{is} i QD_{iinb} Industrial demand

QD_{iagr}

QTR_{ij} Agricultural demand

Export of fresh water
to other districts

B. Recycled water continuity

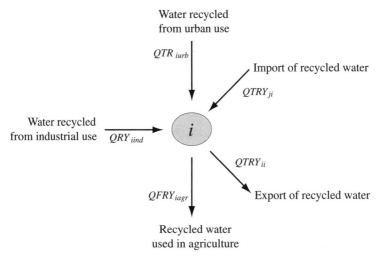

Water recycled
from urban use

QTR_{iurb}

Import of recycled water

$QTRY_{ji}$

Water recycled
from industrial use QRY_{iind} i

$QTRY_{ii}$

$QFRY_{iagr}$ Export of recycled water

Recycled water
used in agriculture

Figure 2.A.1. Continuity of Flows at Node *i*

Crop Choice and Agricultural Demand for Water: Agricultural Submodel

Agriculture in arid and semiarid zones, such as the Middle East, requires water for irrigation. It competes for that water with the household and industry sectors. There are substantial differences in the characteristics of water consumption between the various sectors. For example, compared with agriculture, water demands by households are not price sensitive, at least for the high-priority uses necessary for human life. On the other hand, whereas agriculture can use low-quality water types (recycled, brackish, and untreated surface water), the household sector and some of the industrial sector require mainly freshwater. Another significant difference is that water supply to households and industry must be extremely reliable, whereas the reliance of the agricultural sector on a dependable supply of water may not be as important, especially when water is to be used for low-cash field crops. As a result, agriculture, although the main water-consuming sector, tends to be the one most vulnerable to water shortages. Further, privately profitable, unsubsidized agriculture tends to require relatively low-cost water. This has led to considerable subsidization of water for agriculture.

Although agriculture suffers considerable uncertainty as to water supply, it also has considerable flexibility, since farmers can often produce a large variety of crops and conduct other water-consuming activities (e.g., fish ponds) in the same area. This flexibility is mainly due to annual field crops (e.g., wheat, maize, and cotton) that can be grown using different amounts and qualities of water during different growing seasons. Agricultural planning methods to deal with such issues have been developed and used (Amir

This chapter is largely based on Amir and Fisher (1999). Reprinted with permission from Elsevier.

et al. 1991, 1992), but the sensitivity of agriculture to water remains an important issue for many countries in formulating water policies.

Agricultural demand for water is thus a very important factor in the Water Economics Project; this naturally requires a detailed treatment of demand by agriculture. For that purpose, we have developed a model of agricultural response to water prices and policies, modeling the crop-choice decision of farmers. We believe the mode of analysis is also of interest outside the Middle East. The Agricultural Submodel (AGSM) has two main goals:

- to provide district- and national-level planners with a decision support tool for planning agricultural production under various water amounts, qualities, and prices; and
- to provide the main model with a soundly based analysis of agricultural water demand.

As explained in detail below, AGSM is an optimizing model of agriculture. It uses data on available land, water requirements per unit of land area for different crops, and net revenues per unit of land area generated by the growing of those crops. These net revenues do not include payments for water, which are handled separately. The model takes prices or quantity allocations for water and generates the cropping pattern that maximizes agricultural income. By varying water prices, one can construct demand functions for each water type or for water generally. The model can also be used to examine the effects of water quantity allocations or of nonwater phenomena, such as changes in the prices on agricultural outputs. In the present discussion, however, we concentrate on the demand for water and related matters.

An objection that will naturally occur to the reader has to do with the usefulness of results obtained from an optimizing model. Actual demand curves reflect the behavior of actual people, and actual people may not always respond optimally. To this there are several replies.

- Application of AGSM to actual data suggests that the model closely approximates the actual response of farmers to water prices.
- Even if AGSM generates results that do not exactly agree with actual behavior, those results can serve planners as an approximation. This is likely to prove particularly useful when econometric studies of water demand specific to a district are unavailable.
- A departure of actual behavior from the optima generated by AGSM can serve as a signal to planners that further study should be done.
- AGSM provides a quantitative post–optimal sensitivity analysis that can be used to analyze uncertainty, stability of plans, and risks.
- AGSM can serve as a decision support device suggesting to planners what crop patterns are likely to prove optimal under various conditions and relating these to different water policies.

This chapter presents the basic methods of AGSM, together with some results, but that is not the end of the story. More material on both methodology and results will be found in the appendixes to Chapter 5 and, especially, Chapter 7.

1. The Agricultural Submodel in Detail

AGSM is formulated at the district level. Its objective function is the net agricultural income of the district, which is maximized by selecting the optimal mix of water-consuming activities (crops and fish ponds). In this procedure, the decision variables are the land areas of the activities. Each activity is characterized by its water requirements per dunam[1] and the net income it produces per land area—*not* including water payments. This water-related contribution (WRC) is further explained below. At present, AGSM is a short-term model in the sense that it does not distinguish activities that differ by their capital investment and redevelopment costs. It is also a steady-state model that reflects current data on a yearly basis.

Each activity can, in principle, use one or more types of water: four water quality types and three seasons. As implemented for the Middle East Water Project, the water quality types are fresh (ground) water, surface water, brackish water, and recycled wastewater. There are three seasons: winter, transition (March–April and October–November), and summer. This makes 12 season–quality combinations. AGSM has been formulated to support up to 12 water prices, one for each season–quality and water type.

The constraints in AGSM involve two factors: water and land area. The user can impose constraints on the availability of water by quality and by season. The user does not have to impose such constraints, however; instead, he or she can choose a pricing policy for water, allowing prices to perform any required rationing of supplies. If desired, the user can both have a pricing policy and also specify constraints on water quantities. Different prices can also be set (at the district level) for different water quantities (quotas).

The second set of AGSM constraints involves land areas. These constraints are grouped into categories of the total land area available for agriculture and the total land areas suitable for particular crops and for groups of crops. By specifying land-area constraints in this way, the user can account for the fact that not all land parcels are equally suitable for all activities.

For Israel, for example, the categories of activities subject to land-area constraints (denoted k in equation 3.2.2, below) are as follows:

- all activities;
- all irrigated activities (including fish ponds);
- crops of the same group (field, orchards, flowers, nonirrigated);

- crops irrigated by the same water quality (fresh, recycled, brackish, surface);
- crops grown during the same season;
- unirrigated crops on irrigable land;
- crop rotation; and
- recreation.

The objective function that is maximized in AGSM is the total annual net income of agriculture in the district. Net income is considered in two parts. The first of these is what was referred to above as the water-related contribution. WRC_j, the water-related contribution of activity j, is defined as the gross income generated by activity j per unit area less all direct expenses (machinery, labor, materials, fertilizers) associated with doing so *except for direct payments for water*. WRC_j thus measures the maximal ability of the jth activity to pay for water. WRC enters the objective function positively.

The second component of net income consists of direct payments for water and is subtracted from WRC. It is important to note that such payments do not include water-related expenses such as conveyance and distribution because these are included in the calculation of WRC. This enables us to concentrate directly on the demand functions for water.

Before proceeding, we consider at least two aspects of all this that deserve mention. First, the calculation of WRC does not include any return to farmers on their invested capital. This means that WRC must be considered as the maximal ability of the jth activity to pay for water in the short run. In the long run, farmers will abandon activities that do not recompense them for invested capital. We discuss this issue again below.

Second, because AGSM operates at the district level, the district is treated as a single decisionmaking unit. This means that only one price for each water type (average) is assumed to prevail for all activities within the entire district. One problem associated with such a treatment is that conveyance and distribution costs within the district are, in fact, usually dependent on the location and elevation of water sources and on the location of the agricultural plots on which water from those sources is used. To deal with this limitation, AGSM employs two procedures:

- Where costs vary by water source, we define each source as a separate water type and define activities separately in terms of the source used. Then the different costs will simply show up as different values of WRC for the source-differing activities.
- Where certain activities get water from specific water sources at water prices different from the district average price, we introduce a price correction factor for water consumed by these activities. To illustrate, we take bananas, which are high water consumers: because of climate requirements, banana groves in the Golan district are located near the

Sea of Galilee, 200 meters below sea level, and are irrigated directly from the lake. The average water price in the district, however, reflects the average elevation of the district, which is 100 meters above sea level; the much higher pumping expenses mean that water is supplied at a significantly higher price. In this case, were the high average price of water charged to the banana groves, bananas would leave the optimal basis, distorting significantly the optimal mix of activities. If the district cannot (for other reasons) be disaggregated, then water price correction factors must be introduced.[2]

2. Mathematical Representation of AGSM

The objective function is as follows:

$$Z = \sum_j X_j \left[WRC_j - \sum_i (P_i W_{ij}) \right] \tag{3.2.1}$$

where

WRC_j = as already explained, the water-related contribution of activity j;

P_i = the price of one cubic meter of the ith water type (where i varies by quality and season), $(i = 1, ..., 12)$;

W_{ij} = the demand of water of type i per unit area of activity j; and

X_j = the total land area used by activity j.

Those are the decision variables. The constraints follow.

Land areas

The general form of the area constraints is

$$\sum_j X_{jk} \leq A_k \tag{3.2.2}$$

where

k = the category (see list of categories above);

X_{jk} = the area of activity j in category k (see above); and

A_k = the total area available for category k.

The constraints ensure that the sum of the areas of the crops under each category k will not exceed the area available for that category.

Water

Water constraints are of the following general form:

$$\sum_j W_{ij} X_j \leq W_i \tag{3.2.3}$$

where

W_i = the total available amount of water type i.

Such constraints are formulated for the total amount of water, for seasonal amounts of water, for local water sources, and for water consumed by certain activities.

3. An Example: Applying AGSM to Israeli Data

Detailed results from the application of AGSM to Israel and Jordan are given in later chapters.[3] It will be convenient for our discussion here, however, to consider some examples. We choose examples from Israel, where AGSM was first developed and run.

AGSM was first run on 1994 data for Beit Shean, a district located in the Jordan Valley south of the Sea of Galilee, and then on data for several other districts. We used official data for agricultural production from the Israel Ministry of Agriculture (IMA), Rural Planning Division, and others. Data regarding water were taken from publications of the Water Commission and Mekorot—the national water supply company. Data regarding incomes from crops in all districts were taken from different independent sources (IMA, district planners, Volk 1993, and Israel Farmers Organization). The data were discussed with and approved by the IMA planners of the district.

Calibration runs

To calibrate the model so that it reflects real conditions, we performed some preliminary runs. In these calibration runs we used actual 1994 figures for the right-hand-side (RHS) values of the constraints for water amounts, total land area, and land area of perennial crops (orchards, greenhouse flowers, and fish ponds), because in the short term these land areas are fixed. The land areas for annual crops (winter grains, industrial crops), however, have more flexibility in the short term. Here we permitted deviations up to 15% from the 1994 data. The prices per cubic meter of several types of water in 1994 were as follows: fresh-winter, 12 cents; fresh-summer, 18 cents; brackish, all seasons, 10 cents; recycled, all seasons, 12 cents; surface-winter, 13 cents; and surface-summer, 15 cents.

The outputs of these calibration runs for the optimal land areas, water use, and mix of activities were compared with the corresponding actual 1994 values. Significant deviations between the model outputs and actual data were discussed with the district planners, and this usually led to

Table 3.3.1. Actual and Calculated Data, Beit Shean

Activity	Unit	Actual 1994 data	Model results	Difference (percentage)
Orchards	dunams	14,698	14,500	–1.4
Winter crops	dunams	35,551	35,550	0
Industrial crops	dunams	14,364	14,550	1.5
Total field crops	dunams	49,915	50,000	0
Total vegetables	dunams	19,332	19,040	–1.5
Total unirrigated crops	dunams	11,477	11,500	0
Total cultivated area	dunams	105,570	105,570	0
Total fresh water	cubic meter	54,200,000	45,467,600	–16.2
Total recycled water	cubic meter	2,400,000	2,400,000	0
Total brackish water	cubic meter	35,917,000	35,917,000	0
Total surface water	cubic meter	18,533,000	18,533,000	0
Total water	cubic meter	111,050,000	102,317,600	–7.79

changes in some of the estimated input data (e.g., water-related contribution and water per area unit). When the comparison was satisfactory, we conducted a limited sensitivity analysis, using different water prices and water-related contributions of crops. The model's responses to these changes were compared with the planner's intuitive predictions, resulting in additional changes in model inputs where required. After these runs the model was ready for systematic runs to study the effect of changing water prices. In these runs the calibration restrictions on the RHS of the land-area constraints were removed.

As an example, the results of the calibration runs for the Beit Shean district showed a satisfactory fit to the 1994 actual mix of activities and the use of land and water, as can be seen from Table 3.3.1. In this particular case the model yields optimal total water use as 7.79% less than actual values. Since the model is an optimizing one, this is not a large discrepancy. This was discussed with the planners in the district who checked the reason for the apparent discrepancy. We also carried out a limited systematic set of experiments using different water prices for fresh and saline water. The results regarding both the direction and the magnitude of the changes in production were compared with subjective estimates of the planners and were approved by them.

Obtaining demand curves for water

After the calibration showed acceptable results, AGSM was run systematically to evaluate the response of agricultural production to water prices ranging from 10 cents to $1 per cubic meter. In these runs we used the same water price for all season-quality water types.

In Table 3.3.2, the results for the 14.6-cent water price (second row) are those calculated with all prices set at the weighted average of actual prices for 1994. The results of the systematic increase of water prices for Beit Shean are presented in this table.

The columns of Table 3.3.2 are explained by their respective headings except for the last column. That column shows the activities that leave the optimal basis at different prices as water price increases. For example, the fourth row of the table reads as follows: For P_a = 25 cents per cubic meter, the entire irrigated area is 5,191 hectares, and the total water demand is 38.53 million cubic meters (mcm), resulting in an average of 7,422 cubic meters of water per hectare. Freshwater demand is 18.62 mcm, which is 48.33% of the total amount. Total expenses for water are $9.63 million, and net income is $9.27 million (WRC is the sum, or $18.9 million). Water expenses are 50.95% of WRC. The last column indicates that winter crops cannot pay the price of 25 cents per cubic meter and leave the optimal basis when prices rise from 20 to 25 cents per cubic meter. The last column (of the third line) indicates that the reduction in the irrigated area (from 8,209 to 5,209 hectares is due to the fact that winter crops left the optimal basis.

Columns 1 and 3 in Table 3.3.2 are the water price and the total demand of waters respectively. They are graphically presented in Figure 3.3.1, creating the optimal demand curve for total water for Beit Shean (with all water prices set equal).

In this case the best-fitting curve is of a linear form:

$$P_a = mQ + n \qquad\qquad (3.4.1)$$

where Q and P_a are water quantities and average prices, respectively; and m and n are parameters.

The estimated coefficients for this case are $m = -0.0058$; $n = 0.6098$; and $R^2 = 0.8686$.

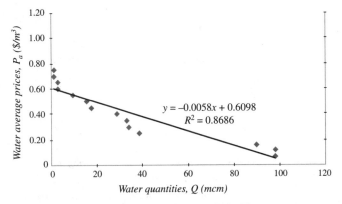

Figure 3.3.1. Optimal Water Demand Curve, Beit Shean

Table 3.3.2. Optimal selected Values for Different Average Water Prices, Beit Shean

Water price, P_a ($/m^3$)	Irrigated area (ha)	Total water use (mcm)	Water per land area (mcm/ha)	Freshwater (mcm)	Freshwater (percentage)	Water expenses ($million)	Net income ($million)	Water expenses per net income (percentage)	Activities leaving the optimal basis as price rises
0.10	8,209	97.76	11,909	23.41	23.95	5.98	22.85	20.74	
0.146	8,209	97.76	11,909	23.41	23.95	10.86	17.99	37.64	Calibration run
0.20	5,209	89.81	17,241	23.41	26.07	14.16	13.33	51.51	Winter crops
0.25	5,191	38.53	7,422	18.62	48.33	9.63	9.27	50.95	
0.30	4,416	34.19	7,742	15.11	44.19	10.26	7.40	58.10	Fish ponds
0.35	4,396	34.01	7,737	15.14	44.52	11.93	5.73	67.55	Maize
0.40	3,139	28.81	9,178	11.99	41.62	11.52	5.73	66.78	Part of orchards
0.45	1,961	17.52	8,934	7.08	40.41	7.89	3.01	72.39	Sunflowers
0.50	1,954	17.45	8,930	6.81	39.03	8.73	2.17	80.09	Part of vegetables
0.55	984	9.50	9,654	6.89	72.53	5.22	1.53	77.33	Spices
0.60	869	8.58	9,873	6.33	73.78	5.15	1.11	82.27	Potatoes
0.65	394	2.88	7,310	1.19	41.32	1.87	0.94	66.55	Vegetables
0.70	394	2.88	7,310	1.19	41.32	2.02	0.85	70.38	Orchards
0.75	0	0		0		0	0.00		All irrigated activities

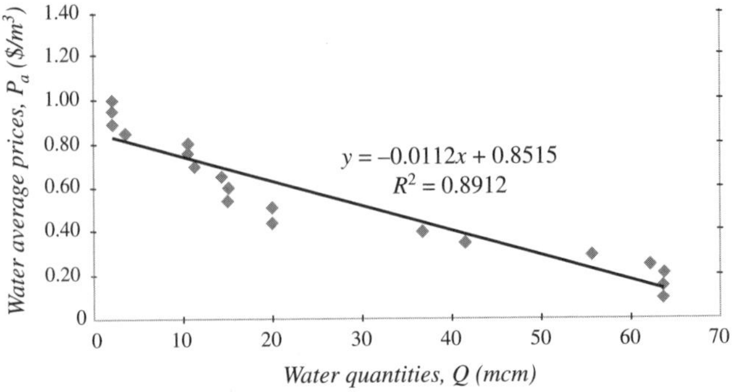

The chart shows data points with a linear trend line. Y-axis: "Water average prices, P_a ($/m^3$)" ranging from 0 to 1.40. X-axis: "Water quantities, Q (mcm)" ranging from 0 to 70.

$$y = -0.0112x + 0.8515$$
$$R^2 = 0.8912$$

Figure 3.3.2. Optimal Water Demand Curve, Hula

Of course, as expected, different districts have different demand curves, as in Figure 3.3.2 for the Hula district.

Detailed results for other Israeli districts are given in the Appendix to Chapter 5. In general, as for Beit Shean and Hula, a linear demand curve fits well. The elasticities of total demand for water for such curves at an average price of 20 cents per cubic meter are reasonable, varying from −0.186 (Merom Hagalil) to −0.488 (Beit Shean), with an average of −0.246.[4]

It should be noted that these elasticities cannot be directly compared with elasticity estimates for water demand by particular groups of crops (e.g., Eckstein and Fishelson 1994). This is because elasticity estimates from AGSM reflect the effects of water prices on competition among crops for limited land. They are therefore affected by the entry into and exit from the optimal basis of particular crops. To put it another way, the elasticity of water demand for particular crops or groups of crops is likely to be greater in absolute value than those for the district as a whole obtained from AGSM. The estimate for a group of crops simply reflects a decrease in water demand when that group becomes unprofitable. The estimates from AGSM include the positive effects on water demand of increases in other crops.

Examination of the optimal demand curves for the various districts shows the following:

- Generally, the generated demand curves have a reasonably regular appearance. The exceptions are for districts with relatively limited agriculture or with a dominant crop, as can be seen in Figure 3.3.3 for the Maale Hagalil district. (See below for further explanations of the jumps in the curves.)
- In several cases (as in Figures 3.3.2 and 3.3.3) the demand curve becomes vertical at the right-hand side. This reflects the fact that, in those districts, at low enough water prices, all the available land area is being used, the

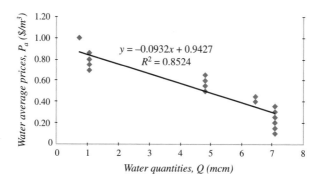

Figure 3.3.3. Optimal Water Demand Curve, Maale Hagalil

optimal mix of activities remains the same, and therefore water demand does not increase as prices drop further. In effect, water demand has an upper limit imposed by other constraints (here, land).

- In many cases the optimal demand curve becomes vertical at the left-hand side (Figures 3.3.1–3.3.3). This reflects the fact that at the higher end of the price range used, irrigated crops become unprofitable; however, crops of very high value, usually of limited area, stay in the optimal basis over the range of prices examined.

So far we have discussed the total demand for water. It is also possible to produce demand curves for specific water types, but these naturally depend on the prices assumed for the other water types. We can, however, gain some insight into the behavior of the demand for particular water types by examining the results for total water. For example, examination of the results for Beit Shean, given in Table 3.3.2, yields the following observations about the demand for fresh (ground) water in that district:

- Freshwater, as a component of the water mix, is used more as prices rise (24% freshwater at P_a = 10 cents and 74% at P_a = 60 cents);
- Both the total amount of freshwater demanded and the average water demand per land unit remain roughly constant for water prices between 45 cents and 60 cents per cubic meter (at approximately 6.3 mcm to 7 mcm and 9,000 cubic meters per hectare, respectively.)

(See Figure 3.3.4, which graphs freshwater demand in Beit Shean as a function of average water price when the prices for all water types are the same.)

Those phenomena can be explained by the fact that the most profitable activities in Beit Shean (as is generally true elsewhere in Israel) are growing vegetables, orchards, and flowers. Of these, vegetables and orchards must be irrigated with freshwater because of health regulations. When water

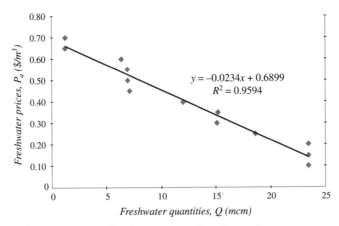

Figure 3.3.4. Demand for freshwater as a function of average water price, Beit Shean

prices increase, the net incomes of all activities decrease until less-profitable activities generate losses and leave the optimal basis. As a result, the amount of irrigated land shrinks (from 8,209 hectares at P_a = 10 cents to just 869 hectares at P_a = 60 cents); only the highly profitable activities that use freshwater remain. Thus, the total amount of water used is reduced with the reduction of the irrigated land, but at the same time, the share of the freshwater component increases because most of the remaining crops are to be irrigated by freshwater.

4. Discussion: Benefits and Problems

Evidently, AGSM is a powerful tool for analyzing cropping decisions and the agricultural demand for water. Further, the user interface developed by the project makes AGSM both flexible and easy to use. But there are both benefits and problems associated with AGSM. We now discuss some of these.

Independent uses of AGSM

Much of our emphasis here has been on the use of AGSM to generate the agricultural demand curve (more correctly, demand functions as explained below) for water. Such curves can be used in the main WAS model. But AGSM has its own, independent interest. That interest stems partly from the ability of AGSM to produce results about specific crops or crop groups and partly from the fact that water prices and policies are not the only inputs that affect crop choice and water demand.

An obvious and important example concerns crop prices. These are used in AGSM as part of the input to WRC. But crop prices themselves are subject to change. In particular, where crops are widely traded internationally, a change in world prices will be reflected in the studied region. There is a serious interest in knowing how domestic or regional agriculture will respond to such changes—an interest that is present whether or not one focuses on water policy. AGSM provides a tool for the study of such effects (and similarly the effects of changes in the prices of nonwater agricultural inputs such as labor).

Integrating AGSM into WAS

Returning to the question of using AGSM to provide appropriate information on the demand for water by agriculture, there is more here than may at first be apparent. Partly because of the difficulties discussed below, one may feel reluctant to tie WAS very tightly to AGSM, using AGSM rather to estimate demand elasticities that in turn are used in WAS. But when one examines the entire set of questions involved, the matter is not so simple. (This was a source of considerable controversy during Phase 1 of the project.) Several points must be considered.

The future of regional agriculture is subject to very considerable uncertainty. Apart from water availability, it is difficult to predict crop prices and also agricultural technology, future water requirements, and crop yields. Will not the use of AGSM, which gives detailed results, produce a spurious kind of certainty since the results are likely to become outdated?

The answer here appears to be that the uncertainties involved are not a result of the use of AGSM but apply to any method of studying future agricultural demand for water. AGSM provides a method of systematically thinking about the effects of various phenomena on crop choice and water demand. The fact that there is substantial uncertainty is not a reason to abandon such systematic thinking. Indeed, it may be a reason to embrace it and to use AGSM to study the effects of uncertainty on farmers. Certainly, as time progresses and factors change, one will wish to reexamine conclusions previously reached, but this is no reason for refusing to reach them at all when the future must be considered in current policy planning.

The notion that one can work in terms of a single price elasticity for water demand or any other very simple formulation is simply incorrect once we introduce more than one water variety. As explained in Chapter 2, we do this in WAS 3.3 by assuming perfect substitutability in agriculture between potable water and treated wastewater at a fixed price difference, but this is not a very realistic treatment. Moreover (as also explained in Chapter 2), when there are several types of water, the demand for each cannot be expressed as a simple demand curve but rather must be analyzed as a demand function into which the prices of all water types enter. Moreover, economic analysis places restrictions on the parameters of such functions, many of which relate parameters in different functions to each other.

When one considers the number of water types that are actually used in agriculture and remembers that future models will have to incorporate seasonal (let alone interyear) effects, the number of separate demand functions becomes quickly very large. In the current version of AGSM, there are 12 such functions (4 water qualities in each of 3 seasons). Hence there are also 12 water prices with which to deal. The available data simply do not come close to permitting the econometric estimation of all these functions, particularly at the district level. AGSM, however, because it is an optimizing model, produces (implicit) demand functions that must satisfy all the restrictions that come from economic analysis.

As a result, we plan to incorporate AGSM directly into WAS in future models in which seasonal effects and multiple water qualities are included. Fortunately, at least as a technical matter, this is easily done by including WRC directly in the objective function to be maximized.[5]

Long-run versus short-run effects

AGSM as employed in our project is a short-term model. As already observed, the values of WRC that are used make no provision for the return on farmers' investment in land or capital equipment. Further, some crops, such as orchards, require care (including water) over several years. Fortunately, there is no reason of principle why AGSM cannot be run as a long-term model making provision for such phenomena. Doing so would also facilitate the study of farmers' response to uncertainty. This is an area for additional research.

Before leaving this discussion, however, we must note a phenomenon of potential long-run social importance already visible in the short-run results—namely, the shrinkage of irrigated land at high water prices. In Beit Shean, for example, winter crops cannot pay for any type of water at a price higher than 20 cents. When prices of all water types increase to 35 cents, fish ponds become unprofitable, and field crops (maize) will not be grown. Because winter crops, fish ponds, and maize currently occupy approximately 50% of the irrigated area in the district, this means that a large area will not be irrigated at such a price. A partial alternative consists of the growing of unirrigated winter crops. However, this alternative is both limited and risky. It is limited to rainy districts and to winter crops only and does not provide an alternative to fish ponds and field crops such as maize. Furthermore, it is risky in years of partial or full droughts.

The inability of much of agriculture to pay high water prices may have undesirable social impacts. For example, economic losses may lead farmers to leave their districts for more industrialized, more populated centers. Such a shrinkage of the agricultural sector would impose difficulties in many countries but be particularly severe for the three entities studied. In Jordan and Palestine, a decline of agriculture would cause much unemployment and great social upheaval. In Israel, where agricultural settlements, especially in districts along the borders (like Beit Shean) have historically been very important for security reasons, and where agriculture is of great ideological importance, such a decline would also mean a very difficult social transition. Whether to avoid such difficulties by subsidizing water for agriculture or other means is of course a matter for national policy—the effects and costs of which can be investigated using WAS.

Constant prices and feedback effects

AGSM, in its present form, assumes that farmers take the prices of agricultural outputs as given. (More generally, it assumes that WRC is constant and independent of the optimizing choices of farmers.) But although this can very well be true of a small, individual farmer, it is much less likely to be true at the district let alone the national level. At such scales, the choices of individual farmers, each of whom takes output prices as given, can very well produce a change in supply in response to water policy that is large enough to influence crop prices. This, in turn, when realized by farmers, will lead to a change in their predicted actions.

An example may assist us here. Suppose that a particular change in water policy causes every farmer to switch from carrots to eggplants, given the prices of all crops. If enough farmers do this, there can very well be a rise in the price of carrots and a fall in the price of eggplants. This, in turn, will moderate the switch.

A related issue involves the discontinuities discussed below and the linear programming nature of the model. Without the imposition of outside constraints, the running of AGSM at constant crop prices could produce a result in which the production of some particular crop is so large as to be unreasonable. Although we have not encountered such a case in practice, the possibility is certainly there. One way to prevent it would be to model the effect of cropping decisions on agricultural prices.

Unfortunately, this is not a simple matter. To do it would require modeling the demand functions for agricultural outputs, and this would mean expanding an already enormous enterprise. Moreover, a similar situation could arise in the analysis of industrial demand for water. At least for the present, we have not dealt directly with such problems, but they may require handling in future work.

Discontinuities in the demand curves

It will be apparent from examination of Figures 3.3.1–3.3.4 that AGSM can produce jumps in the demand for water as prices change. (Look at the dots, not the lines fitted to them.) This phenomenon can occur as an artifact of the method and assumptions used in AGSM, or it can happen for "real" reasons when a district has only a limited number of possible water-consuming activities. We now consider both possibilities.

1. Artifact. Consider Figure 3.3.4 for Beit Shean. The curve appears discontinuous when water prices increase from 45 to 60 cents per cubic meter while freshwater quantities remain stable at about $Q = 7$ mcm. This apparent discontinuity is due to a high water-related contribution crop that remains in the optimal solution for water prices smaller than 60 cents per cubic meter. The linear programming nature of the model, together with the fact that all plots of land using a particular activity are assumed to be identical, means that when water prices increase, other things being equal, the model assumes that there is a single "critical" price, P_j^*, at which the net income of a particular activity j, N_j, becomes suddenly zero on all plots.

$$N_j \equiv WRC_j - W^j P_j^* = 0 \qquad (3.4.1)$$

where W^j denotes the total water used for activity j and we continue to assume (for simplicity) that all water types have the same price.

Now consider the results for freshwater in Beit Shean (Figure 3.3.4). Since the critical prices, P_j^*, for flowers, vegetables, and orchards are all higher than 60 cents, these crops remain in the optimal basis when the price rises to 60 cents. At a price of 65 cents, however, orchards leave the basis, cultivated land decreases to 394 hectares, and freshwater demand is sharply reduced, from 6.33 mcm to 2.88 mcm (for flowers). In reality, however, such a reduction would be gradual because not all plots are identical, and therefore, orchards would leave the basis in a continuous fashion, smoothing the demand curve. A method for smoothing such discontinuities is presented in the appendix to this chapter. Of course, the linear nature of AGSM is not a matter of modeling principle.

2. Real jumps. Figure 3.4.3 illustrates the second reason for jumps in the demand curve—namely, the presence of only a limited number of activities in the district. In Maale Hagalil there are three main crops. In this case the jumps appear larger than can be explained by statistical distributions of the WRCs and, apparently, reflect a real situation.

Appendix: Smoothing Agricultural Demand Curves

As discussed in the text, discontinuities in the curves can be due to the assumption that all plots of an activity are the same. In this respect, a smooth curve may better reflect real conditions. The following method for

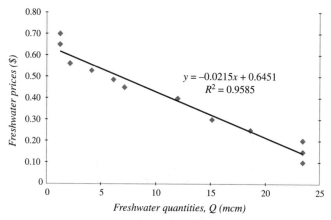

Figure 3.A.1. Smoothed Freshwater Optimal Curve, Beit Shean

smoothing curves has been developed. We present this method in terms of its application to freshwater in Beit Shean.

Usually, each activity in a district involves many plots. Each of these plots is operated under different conditions, resulting in a different water-related contribution. Because of the large number of plots in a district, it makes sense to assume that the water-related contribution of the jth activity, WRC_j, is normally distributed. Referring to equation 3.4.1, $WRC_j = W^j P_j^*$, and, since W^j is not stochastic, the critical price, P_j^*, is also normally distributed with mean, μ, and standard deviation σ, say.

Since the range $\mu \pm 2\sigma$ includes more than 95% of the values of the distribution, we assume that the range of discontinuity in Figure 3.4.4 (between $P_a = \$0.45$ and $\$0.60$) equals $\mu \pm 2\sigma$, where $\mu = (0.45 + 0.60)/2$. We divide that range into four equal segments, each of which is of length $\sigma = (0.60 - 0.45)/4$, where $P_a = \$0.45 = \mu - 2\sigma$ and $P_a = \$0.60 = \mu + 2\sigma$. According to the normal distribution, approximately 16%, 50%, and 84% of the plots will generate zero net income at the critical prices $P_j^* = \mu - \sigma = \$0.4875$, $\mu = \$0.5250$, and $\mu + \sigma = 0.5625$, respectively. For each of these new critical prices we can estimate the demand for water by the remaining plots.

The resulting quantities are for $P_a = \mu - \sigma = \$0.4875$, $Q = 7.08 - 0.16 \times (7.08 - 1.19) = 6.14$ mcm; for the mean price $P_a = \$0.525$, $Q = 7.08 - 0.50 \times 5.89 = 4.14$ mcm; and for $P_a = \$0.5625$, $Q = 7.08 - 0.84 \times 7.08 = 2.13$ mcm. The "smoothed" freshwater optimal curve is presented in Figure 3.A.1.

4

International Conflicts: Promoting Cooperation

As described in the Preface, water is usually considered in terms of quantities only. Demands for water are projected, supplies estimated, and a balance struck. Where that balance shows a shortage, alarms are sounded and engineering or political solutions to secure additional sources are sought.

Disputes over water are also generally thought of in a similar way. Two or more parties with claims to the same water sources are seen as playing a zero-sum game. The water that one party gets is simply not available to the others, so that one party's gain is seen as the other parties' loss. This is true whether the parties are different countries, different states or regions, or different consumer types.

But as we have seen, there is another way of thinking about water problems. That way involves thinking about the economics of water and, especially, about the value of water in different locations. As already shown, this leads to powerful tools for water management and the cost–benefit analysis of infrastructure projects. We shall now see that it also leads to ways of resolving water disputes and promoting cooperation. By doing so, what appears to be a zero-sum game can be turned into a "win–win" situation.

The fact that (save in landlocked countries) desalination puts an upper bound to the value of water in dispute is dramatic and easily understood. It means, for example, that the value of the water in dispute between Israelis and Palestinians is considerably less than $100 million per year. Such amounts ought not to be a bar to agreement between nations.[1]

To see what is involved here, we return to the discussion begun in Chapter 1, repeating some of it for convenience.

1. Water Ownership and the Value of Water

We have seen that water can be viewed as an economic, if special, commodity. That view has at least two implications for the design of a lasting water arrangement that is to form part of a peaceful agreement among neighbors. The first has to do with negotiations over the ownership of water quantities. The second and, we believe, more important implication has to do with the form that a water agreement should take.

Two basic questions are involved in thinking about water agreements: the question of water *ownership* and the question of water *usage*. One must be careful to distinguish them.

All water users are buyers, in effect—whether they own the water themselves or purchase it from another party. An entity that owns its water resources and uses them itself incurs an opportunity cost equal to the amount of money it could otherwise have earned through selling the water. An owner will use a given amount of its water if and only if it values that use at least as much as the money to be gained from selling.[2] The decision of such an owner does not differ from that of an entity that does not own its water and must consider buying needed quantities of water: the nonowner will decide to buy if and only if it values the water at least as much as the money involved in the purchase. Ownership only determines who receives the money (or the equivalent compensation) that the water represents.

Water ownership is thus a property right entitling the owner to the economic value of the water. Hence a dispute over water ownership can be translated into a dispute over the right to monetary compensation for the water involved.

The property rights issue of water ownership and the essential issue of water usage are analytically independent. For example, resolving the question of where water should be efficiently pumped does not depend on who owns the property. Although both issues must be properly addressed in an agreement, they can and should be analyzed separately.[3]

The fact that water ownership is a matter of money can be brought home in a different way. It is common for a country to regard water as essential to its security because water is required for agriculture and countries wish to be self-sufficient in their food supply. This may or may not be a sensible goal, but the possibility of desalination implies that every country with a seacoast can have as much water as it wants if it chooses to spend the money to do so. Hence, so far as water is concerned, every country with a seacoast can be self-sufficient in its food supply if it is willing to incur the costs of acquiring the necessary water. As a result, disputes over water among such countries are merely disputes over costs, not over life and death.

Of course, self-sufficiency in agriculture can be quite expensive. That makes naturally occurring water more valuable than would otherwise be the case. But this water cannot be worth more than the cost at which it could be replaced by desalination. Indeed, it is typically worth less, since there are costs associated with naturally occurring water as well.

Now, the fact that disputes over water can be expressed as disputes over money may be of some assistance in resolving them (although as we shall see, this is not the principal point for the analysis of and the benefits from cooperation).

Consider bilateral negotiations between two countries, A and B and different proposed allocations of ownership rights between them. Each of the two countries can use its Water Allocation System tool to investigate the consequences to it (and if data permit, to the other) of the allocation. This should help it in deciding what terms for which to settle, possibly trading off water for other, nonwater concessions. Indeed, if at a particular proposed allocation, A would value additional water more highly than B, then both A and B could benefit by having A get more water and giving B other things that it values more. Note that this does *not* mean that the richer country gets more water. That happens only if it is to the poorer country's benefit to agree.

Of course, the positions of the parties will not be expressed along those lines. Their positions will run in terms of ownership rights and international law. The use of the methods here described in no way limits such positions. Indeed, the principal point of this section is not that the model can be used to help decide how allocations of property rights should be made. Rather, the point is that water can be traded off for nonwater concessions. The WAS tool provides a way of measuring such trade-offs.

Moreover, the trade-offs need not be large. Recall that desalination puts an upper bound on the value of water in dispute. Moreover, because naturally occurring freshwater must be pumped, treated, and transported, the upper bound on the value of a cubic meter of such water in situ will be considerably less than the cost of desalination per cubic meter. At the limit (in this example), 100 million cubic meters (mcm) annually of the disputed water of the Mountain Aquifer (a large amount of water in the Israeli–Palestinian dispute) can never be worth more than roughly $20 million per year,[4] and the empirical results of our project show that in fact, the value is far less even than this. Moreover, we use 100 mcm only as indicative. The conclusion would be the same if we used 200 mcm, 300 mcm, or more.

Such sums are small relative to most gross domestic products. They are certainly small relative to the cost of modern military equipment. By monetizing water conflicts, such conflicts can cease to seem insoluble.

A specific example will help illustrate those points.[5]

Water on the Golan Heights is often said to be a major problem in negotiations between Israel and Syria.[6] By running the model with different amounts of water, this question can be evaluated. We have done so. In 2010, the loss of an amount of water roughly equivalent to the entire flow of the Banias springs (125 mcm annually) would be worth no more than $5 million to Israel in a year of normal rainfall and only around $40 million in a year with a 30% drought. At worst, water can be replaced through desalination, and thus this water (which has its own costs) can never be worth more than about $75 million per year.

Those results take into account Israeli policies toward agriculture.

It is *not* suggested that giving up so large an amount of water is an appropriate negotiating outcome, but water is not an issue that should hold up a peace agreement. These are trivial sums compared with the Israeli GDP or the cost of fighter planes.[7] Moreover, a similar analysis applies to the Hasbani, which seemed a possible *casus belli* when Lebanon proposed to pump substantial amounts of water a few years ago.[8]

The preceding example and discussion point to two ways in which the use of WAS models may be helpful in negotiations:

- By expressing water ownership in monetary terms, parties may bring themselves to consider trading off ownership claims for other, nonwater benefits. As the example shows, those trade-offs need not be large.
- Each party to the negotiations can use a WAS model for its own water economy to investigate the consequences of different water agreements.

The former point may very well be politically difficult. Water is an emotional topic, and countries—and their bodies politic—may consent to give up water for other concessions only after an extensive campaign of public education, if at all. Such education may be made easier and the same aim accomplished if the matter involves not trading away dearly held (or at least dearly claimed) ownership rights but rather cooperation in water. This is a very important issue, and we discuss it at length in the next section.

In any event, whether or not water agreements are regarded as solely matters of economics (broadly defined), they surely have economic consequences. Each party can inform its own negotiating position by using a WAS model to investigate those consequences. At the very least, such models can assist by discovering whether some part of the "parade of horribles" that may be thought to be consequent on compromise in water has any basis in fact.

2. Cooperation: Gains from Trade in Water Permits

As just suggested, using WAS models in negotiation is only the beginning of the story, and in fact, there is a good deal more to be said. The simple and final allocation of water quantities in which each party uses what it "owns" is not an optimal design for a water agreement. As we shall see, it is possible to improve on such a fixed-quantity agreement, and the potential gains from doing so can be so large for all parties as to make the question of water property rights a matter largely of symbolic significance.

As we have seen, efficient allocation of water simulates a market solution. In such a solution, if shadow values in two locations differ by more than the cost of conveyance, then there are gains to be had from conveying water from one location to the other. That is true even if the two locations are inhabited by citizens of different countries. Hence, a tool such as the WAS model can serve as a guide for water allocation not only within a country but also among countries.

How would this work? Suppose for the moment that property rights issues have been resolved. Since, as we have seen, the question of water ownership and the question of water usage are analytically independent, it will generally not be the case that it is optimal for each party just to use its own water. Instead, consider a system of trade in *water permits*—short-term licenses to use each other's water. No sale of sovereign rights would be involved. The purchase and sale of such permits would be in quantities and at prices given by an improved and agreed-on version of our optimizing model.[9]

Let us see whether there would be mutual advantages from such a system and whether the economic gains might be a natural source of funding for water-related infrastructure. First, consider the fact that both parties to a voluntary trade gain. The seller would not sell unless it valued the money received more than the water given up; the buyer would not buy unless it valued the water obtained more than the money it paid. Although it is true that one party may gain more than the other, such a trade is not a zero-sum game but rather a win–win opportunity. Moreover, the fact that such trades would take place at model-produced prices would preclude monopolistic exploitation. Indeed, particularly if cooperative infrastructure is built to facilitate trade, the gains from cooperation in this matter appear so large as to dwarf the value of ownership transfer of reasonable amounts of water.

We postpone detailed discussion to Chapter 8, but some of the results obtained with WAS for Israeli–Palestinian cooperation can be summarized as follows:

First, cooperation is a win–win policy that can be worth as much as $100 million per year by 2010. It is far more valuable than are any likely changes in the ownership of the water itself. Although the exact gains from cooperation naturally depend on the assumed allocation of ownership rights, both parties would always gain from cooperation. Note, in particular, that the gains to the selling party are over and above the amounts necessary to compensate its consumers for higher priced or less water. In plausible runs, we find water permit sales going in both directions, depending on the geographic distribution of ownership rights, demands, and infrastructure, especially conveyance systems. With cooperation, the value of the entire Mountain Aquifer will be considerably less than $100 million per year in 2010. The value of the possible differences between the parties' ownership claims will be far less even than that.

Second, desalination on the Mediterranean coast will not be needed in normal years. With cooperation in water and the construction of infrastructure (recycling plants and conveyance systems, largely for the Palestinians), by 2010, additional sources of water will be needed only in years of considerable drought.

Third, the need for desalination will crucially depend on the status of cooperation in water, however. Without such cooperation and with the 1995 ownership allocations, the Palestinians will find desalination at Gaza attractive by 2010.[10]

Finally, the construction of recycling plants in the West Bank, and, particularly in Gaza, will be highly beneficial regardless of water ownership or cooperation. Among the gains that would arise from cooperation and joint infrastructure is the following: the model strongly suggests that even in the presence of current Israeli plans, it would be efficient to have a water treatment plant in Gaza with treated effluent sold to Israel for agricultural use in the Negev, where there is no aquifer to pollute. (Indeed, since this suggestion arose in model results, there has been discussion of this possibility.) Both parties would gain from such an arrangement. This means that Israel has an economic interest in assisting with the construction of a Gazan treatment plant. This possibility—a serious act of cooperation—and other potential projects would, of course, have to be carefully evaluated. But there can be little doubt but that valuable joint projects benefiting all parties can be located and built.

Beyond pure economics, moreover, the parties to a water agreement would have much to gain from an arrangement of trade in water permits. Water quantity allocations that appear adequate at one time may not be so at other times. As populations and economies grow and change, fixed water quantities can become woefully inappropriate and if not properly readjusted, can produce hardship. A system of voluntary trade in water permits would be a mechanism for flexibly adjusting water allocations to the benefit of all parties and thereby for avoiding the potentially destabilizing effect of a fixed water quantity arrangement on a peace agreement. It is not optimal for any party to bind itself to an arrangement whereby it can neither buy nor sell permits to use water.

3. Possible Objections

We now discuss several objections that may be raised in principle to such a plan.

Money cannot buy water

The first possible objection is that it is offensive to suppose that historic water rights can or should be traded for money. This is an objection of form rather than one of substance.

In the first place, the system of trade suggested would not in fact trade sovereign water rights. It would trade short-term permits to use water. Ownership, and hence symbolic control, would not be traded. Second, the trade need not be for money itself. Rather, it makes sense that short-term water permits should be granted in exchange for infrastructure development that benefits all parties or simply benefits the party granting the water permits. Such an exchange can be thought of as water-for-water, at least in the long run. Money is only the way one keeps score.

Settling property rights: An interim escrow fund

The second objection is that the system here described does not settle the property rights issue. Indeed, it does not pretend to do so, although this way of thinking about water should make negotiations more tractable. But does not the institution of trade in water permits and cooperation in infrastructure require that property rights be first settled?

The answer to this is no, although settlement of property rights issues is very desirable. While property rights negotiations are proceeding, trade in water permits could begin with payments made into an escrow fund. That fund would be jointly managed and provide a source of financing for mutually desirable infrastructure. Negotiations over water property rights would effectively become negotiations over shares of or obligations to the fund plus entitlements to future payments. This is as it should be, since water property rights are a matter of money.

Because the gains from trade in water permits can be large relative to the value of water property rights themselves, it would be foolish to wait to reap the benefits from this trade because it is difficult to settle a matter of relatively small monetary magnitude.

Commitments cannot be made under uncertainty

A third possible problem is that unforeseen events, such as droughts, might render a commitment to sell at model prices harmful. There is a two-fold answer here.

First, although the present WAS model is for a single year, it appears entirely possible to build a multiyear model and incorporate the effects of climatic uncertainty. Even in a single-year model, however, repeated runs can yield information on the value of water in unusually dry or wet situations.

Second, the user can place a positive value on the retention of a reserve even without a precise estimate of such value. This would form part of the social value and then be incorporated into the prices at which sales take place. Recall that only *willing* sales (and purchases) are involved. Nobody is forced to sell.

National values would be compromised

A fourth possible objection has to do with the consequences of committing to the use of such tools in a regional context. Does not the user give up data security? What happens to domestic water policies and national values?

It is the latter issue that appears the more important one. Data, in the sense of data on actual water supplies and actual consumption, cannot (or ought not to be) very sensitive. No agreement of any sort is likely to be possible without an agreement on the facts.

The right of each country to set its own national policies toward water, however, should not be questioned. But the WAS tool permits such policies to be set and examined and rethought. Given those policies, the model can then be used to support trade in water permits. Any sort of cooperation must take such policies into account.

Furthermore, we have obtained a result that may seem surprising at first glance. Consider the situation of two countries, A and B, trading in water permits as described. Suppose that A now chooses to subsidize water for agriculture. This will apparently have two effects on B. First, the output effect: if A's agriculture competes with B's, this will give A's agriculture an advantage. Second, the water effect: the increased demand for water in A as a consequence of the subsidy will raise the shadow value of water in both countries, and this will disadvantage B and its consumers. An agreement to trade in water permits, the argument goes, would necessarily lead to constant negotiations over what domestic water pricing policies can be permitted.

About the output effect, we can only say that A could also give its agriculture a competitive advantage through a direct subsidy. Hence, this is not really a matter of water policy even though the result may be brought about through water.

In the case of the water effect, the situation is not as it appears. The subsidy-induced increased demand in A will indeed raise water shadow values in B. (In what follows, we assume that B's consumers are charged those shadow values. To deal with other cases would only complicate the exposition without changing the basic results.) Consumers in B pay higher prices, reducing the benefits they obtain from water. Some of that loss is simply a greater payment to (public or private) water sellers in B itself. As such, it is a transfer within B and not a loss to B as a whole. The remaining loss (called by economists a deadweight loss) is a loss to B as a collective entity;

it involves the fact that B's water consumers reduce their water consumption as a result of higher prices.

But there is also a third effect. If A is importing water from B, then it will be paying higher prices for those imports as a result of its own subsidy. This is a net gain to B and one that can be used to compensate B's consumers. Call this the international trade effect.

As reported in Chapter 8, we have experimented in the case of Israel and Palestine to see the net results of all this. We find that for any reasonable pattern of ownership, the effects on Israel or Palestine of an agricultural subsidy by the other are either negligible or, because of the international trade effect, slightly positive. Although the costs of subsidy policies can be high, those high costs are borne by the party doing the subsidizing. As a result, within a wide range, an agreement to trade in water permits need not be an agreement to repeatedly negotiate domestic water policies.

The scheme would encourage misrepresentation and gaming

A somewhat related issue concerns the possibility that the parties to voluntary trade in water permits would deliberately misrepresent their demands for or policies toward water so as to gain an advantage. In this connection, note first that a party that acted in this way would run some risk. If a party that is a buyer were to overstate its demand, it would end up paying prices higher than its true value of the water obtained. Similarly, if a party that is a seller were to understate its demand, then it would end up selling water at prices below its true value.

This does not end the matter, however. Since water demand is likely to be inelastic at reasonable prices, a party that is a seller might gain by overstating its demand. In such a case, the selling party would retain some water that it values less than the price, but it might succeed in earning sufficiently greater revenue from the water it does sell to leave it better off. In effect, such a seller would be exercising market power by withholding water from the market and exploiting the fact that it faces a declining (and inelastic) demand curve. (An analogous statement holds for a buying party understating its demand: the supply curve facing the buyer is effectively the demand curve of the seller and is hence also inelastic.) The fact that trade leads to gains shows that there is a surplus to be split among the parties; behavior of the sort described could affect the way in which that surplus is divided.

How important this phenomenon is likely to be may depend in part on the overall atmosphere in which trading in water permits takes place. But such misrepresentation is not likely to be easy or long repeated. We are talking here about misrepresentation either of objective demand data or of policies to be applied. (Misrepresentation of costs can also matter.) These are issues of checkable facts, rather than projections of events long in the future, and parties should be able to agree on how to check them. That includes checking actual water consumption and verifying that announced water policies are actually carried out.

Two more observations are worth making. First, even if such misrepresentations are successful, there will still be a surplus to be divided and both sides will gain relative to a fixed quantity agreement. Second, changing the debate about water rights into a discussion of facts and data would itself be a gain in settling water issues.

National security: Hostages to fortune

The major objection to trade in water permits among previously hostile neighbors, however, is likely to involve security. When an agreement is reached among long-term adversaries, is it wise to rely for water on a promise of trade? What if the water were to be cut off?

There are several points to be made here. First, the geographic situation does not change with an agreement to trade in water permits. Thus, if an upstream riparian country could cut off a downstream neighbor's water in the presence of an agreement, it could equally well do so in its absence.

A system of trade in water permits, however, makes this less likely to happen, because it is a system in which continued cooperation is in the interest of all parties. When joint infrastructure has been constructed and gains from trade in water permits are large, withdrawal from the trade scheme will hurt the withdrawing party.

There is, however, one aspect of reliance on an agreement to trade in water permits that does raise an issue. Where the agreement leads either to the construction of infrastructure that would become useless or to the failure to construct infrastructure that would be needed if trade were cut off, reliance on trade may involve some risk. In effect, in such cases, one or another of the parties may be giving hostages to fortune.

Are such cases likely in the Israeli–Palestinian case? We begin with Israel. If there were to be an agreement with the Palestinians along the lines we have suggested, it would make sense for Israel to invest in trade-facilitating infrastructure. Were trade to cease, that investment would largely be lost. This does not seem a major problem, however.

The reverse problem—failure to build infrastructure that would become vital in the absence of trade in water permits—does not seem at all serious. Israel now has a well-developed infrastructure. There does not appear to be any project that would be both unnecessary in the case of an agreement to trade and vital if such trade were suddenly to cease.

The Palestinians, by contrast, may have more exposure in the form of hostages to fortune. Without trade in water permits, and with an unfavorable agreement on West Bank water property rights, the Palestinians would soon be forced to build desalination plants in Gaza. In the presence of trade, such plants would be unnecessary for a long time to come. Hence, if an Israeli–Palestinian agreement takes the form of trade and cooperation, the Palestinians will have to consider whether they should build such desalination plants in any case. If they do, they will lose a good deal of the economic benefits from trade. If they do not, then there may be a problem should trade cease.

What that choice should be depends on how likely it is that Israel would abrogate such an agreement and on the situation that one believes would then arise. For example, in such an event, presumably the Palestinians would feel justified in extensively pumping the Mountain Aquifer, even if that were not the regionally efficient or agreed-on thing to do. Surprisingly, however, we find that in the absence of cooperation, the Palestinian need for desalination would stem not directly from the need to use desalinated water in Gaza itself but from the need to (inefficiently) supply the southern West Bank from the Gazan desalination plants by piping it uphill to the area of Hebron (see Figure P.1 and Chapter 6). Hence, the apparent crisis caused by an Israeli abrogation of a cooperative water treaty could be overcome by Palestinian pumping of Mountain Aquifer water beyond the amounts permitted by the water treaty, doing so until the needed desalination facilities could be built.[11]

But a principal incentive for the Palestinians to participate in the win–win kind of agreement that we have described must lie in their belief in two other points. First, they must believe that it is very much in Israel's own interest to continue participating in the agreement. Second, they must believe that Israel understands its own interest sufficiently well to abide by the commitments it makes.[12] That kind of trust must be a principal feature of any peace negotiations.

4. Concluding Remarks

We summarize the main points of this chapter. First, careful attention to the economics of water and to the difference between water ownership and water usage leads to the construction of a powerful analytic tool—an optimizing model of the water system or systems. This model can be an important aid to policymakers in their water management and policy decisions.

The usefulness of this approach does not end at the international border, however. A modeling effort and the analysis accompanying it can also be used in the resolution of water disputes. That use has at least two aspects. First, property rights in water are seen to be reducible to monetary values. If this is done, negotiations over water can cease being limited to water itself and be conducted in a larger context in which water is measured against other things. Moreover, the availability of seawater desalination means that the monetary value of disputed water property rights will generally not be very large.[13] If this is realized, negotiations over water should be facilitated.

The approach has another implication of at least equal importance, however. Water agreements that simply divide water quantities are not optimal and may be very bad agreements indeed. Such fixed-quantity agreements are zero-sum games in which the gain of one party is the loss of the others. Instead, it is possible for disputants to engage in a win–win arrangement where permits to use water are traded among them. Especially when such cooperation involves the construction of mutually beneficial infrastructure, the gains to all parties can be quite large, considerably larger than the value of the water property rights themselves.

Moreover, such gains need not only be economic ones. Such cooperative arrangements can provide the kind of flexibility that can keep changing water needs from disrupting a peace agreement. Further, cooperation in water and in water-related infrastructure can be a confidence-building measure. In this way, water can cease to be a source of continued conflict and instead become a source of cooperation and trust.

Part II

Results for Israel, Palestine, and Jordan

A Note on Some
Sensitive Issues

We now come to the presentation of detailed results for Israel, Palestine, and Jordan, as well as results concerning regional cooperation. Before doing so, it is appropriate to discuss some sensitive issues and the way in which we have dealt with them. In this book, we have tried not to take any political position, but some matters cannot be wholly avoided.

The first set of issues concerns nomenclature. Terminology can have considerable symbolic importance. We have already discussed in the Preface our use of the term "Palestine" and the fact that we use different names for the large lake on the Upper Jordan (Kinneret, Lake Tiberias, or Sea of Galilee) depending on whether we are discussing results for a particular country or for the region. Such usage, of course, has no bearing on the substantive results.

More substantive are the conflicting claims as to land (and of course water). We have not attempted to resolve those claims, nor have we forced the positions taken by the various country teams to be mutually consistent. Instead, in each of the country chapters, we take those positions as given.

This presents little difficulty in general. Of course, in Chapter 8 we discuss what difference various outcomes as to water negotiations make, but beyond that, there are two issues that require discussion.

Israeli settlements

Sensitive as this issue is, it creates no real difficulty for the understanding of our results. In the Israel chapter, we assume the water demands of such settlements to be in place. In the Palestine chapter, they are ignored. In the chapter on cooperation, we take the view (implicitly) that under cooperation, either the settlements must be supplied with water or, alternatively, there will be water demands from Palestinians replacing the Israelis. Because the water demands in question are not very large, this has no appreciable effect on our results.

Jerusalem

The Israeli team naturally assumes that the population of their Jerusalem district ("Jerusalem Mountains") includes that of East Jerusalem. The Palestinian team, equally naturally, assumes that the East Jerusalem population is theirs. This means that in both the Israel and the Palestine chapters, that population is supplied by the corresponding water system.

That by itself presents no difficulty. The issue arises only when we look at cooperative runs of the model in Chapter 8. Even there, when we ask how a cooperative system would manage water, we can avoid the problem simply by not double-counting the population in question. When we examine the gains from cooperation, however, we need to compare the net benefits from cooperation with those achieved without it. The latter cannot be estimated without some decision on the responsibility for serving East Jerusalem. Since any such decision by us would be inappropriate, we again ignore the inconsistency and have the population in question doubly served in both cooperative and noncooperative runs. Of course, in the context of an overall peace agreement, the analysis could readily be adjusted. The impact here is small in any case, although the political issue is not.

Finally, we must comment on data quality and on the limitations of the Water Allocation System model. The results are, we believe, interesting and sometimes surprising. But results depend on data. We have done our utmost to find and use the most accurate data available, but since official data on water are sometimes not released, it may well be that our data are not fully accurate. Even if that is so, however, it is not an adequate reason for simply dismissing our results. Those results show what our tools can accomplish. If water authorities believe that our data are inaccurate, we encourage them to use the same tools with the correct data. In any case, we do not think that our data are likely to be so inaccurate that our qualitative conclusions would change. (A similar statement is true of our demand projections for later years. WAS permits the user to examine sensitivity to demand assumptions.)

At present, WAS itself is limited, in that it is a single-year rather than a multiyear model. A multiyear model would permit studying the optimal timing of infrastructure projects and also the interyear management of water resources, particularly aquifers. The development of such a model is a prime candidate for later research, as is a more sophisticated treatment of water quality issues than has so far been possible. But the results from the current version of WAS are powerful, and we do not believe that their presentation need await further model development. As Robert Hall of Stanford University once said, "Research projects are never completed, only abandoned." Our project has certainly not been abandoned and will always be improvable.

Results for Israel

In this chapter, we apply the Water Allocation System model to Israel, deriving results for the value of water under various assumptions and analyzing the costs and benefits of infrastructure projects.

1. Data for 1995[1]

Districts

We divide Israel into 20 districts. (See Figure 5.1.1.) These districts are those used by the Israeli Ministry of Agriculture, and its planning and research teams use these definitions of districts for their data collection and economic planning of agriculture and the related Israeli water system. Known as ecological districts, they are distinguished by their ecological characteristics.[2] Recall that the model uses a single node in each district, at which supply and demand are concentrated, and that a conveyance system connects the districts.

Water consumption data and water demand curves

The source for water consumption data is *The Agricultural Production Forecast* for the years 1995–2020, published by the Ministry of Agriculture, May 1996 (Hebrew). The data are presented in Table 5.1.1. The quantities for urban and industrial sectors refer to freshwater only, while those for agriculture are the sum of freshwater, recycled water, and other nonpotable sources, such as saline and floodwater (total quantity used). The model assumes that fresh and recycled and other nonpotable water are perfect substitutes for farmers.[3]

Figure 5.1.1. Partial Regional Map, Israel

Table 5.1.1. 1995 Water Consumption (mcm)

District	Urban	Industry	Agriculture
		Sector	
Golan	3.50	0.50	44.42
Hula	5.00	1.00	79.97
Merom Hagalil	4.00	0.50	34.79
Maale Hagalil	7.50	1.00	7.12
Acco	48.00	12.50	63.93
Biqaat Kinarot	5.50	0.50	46.19
Beit Shean	3.00	0.00	123.19
Gilboa Harod	1.00	0.00	28.05
Lower Galilee	4.00	1.00	20.62
Jezreel Valley	8.00	1.50	58.55
Nazareth Mtns.	22.00	4.00	12.28
Hadera	28.50	24.00	154.68
Raanana	197.00	19.00	99.04
Rehovot	67.00	17.50	130.10
Jerusalem Mtns.	65.00	6.00	33.22
Lachish	11.50	12.00	72.11
Habsor	8.00	1.00	67.65
Negev	32.50	3.00	100.14
Arava	9.00	3.50	42.82
Jordan Valley settlements	10.50	1.00	34.03
Totals	540.50	109.50	1,252.90

Total demand = 1,902.90

In Israel, the water prices charged to consumers are fixed. We assume that these consumption figures correspond to "base" prices of $1 per cubic meter for urban and industrial sectors and 16.5 cents per cubic meter for agriculture (the actual prices in 1996).[4] The combinations of price and quantity define for each sector and each district a point on the water demand curve.

To obtain the complete demand curve in the relevant range, we assume constant elasticities: –0.2 for the urban sector, –0.33 for industry, and –0.5 for agriculture.[5]

Supply data

Our supply data come principally from the Israel Water Master Plan (Water Commission and TAHAL), September 1996 (Hebrew), and partially from other sources (Mekorot, the Hydrological Service, the Agricultural Planning Center, and other publicly available publications[6]).

Table 5.1.2. 1995 Supply Data

District	Step 1 Quantity (mcm/ year)	Cost (¢/m³)	Step 2 Quantity (mcm/ year)	Cost (¢/m³)	Step 3 Quantity (mcm/ year)	Cost (¢/m³)	Local nonpotable sources (mcm/year)
Golan	5	6.0	30	19.0	25	19.0	6
Hula	150	6.0	50	8.3	—	—	26
Merom Hagalil	—	—	—	—	—	—	4
Maale Hagalil	2	7.0	—	—	—	—	0
Acco	28	5.0	70	7.0	—	—	20
Biqaat Kinarot	662	0.0	—	—	—	—	19
Beit Shean	38	3.0	14	9.0	—	—	71
Gilboa Harod	18	3.0	8	3.0	—	—	5
Lower Galilee	16	25.0	—	—	—	—	1
Jezreel Valley	14	6.0	8	21.0	—	—	13
Nazareth Mtns.	4	8.0	6	25.0	—	—	0
Hadera	61	4.0	55	9.0	—	—	30
Raanana	158	5.0	79	9.0	—	—	1
Rehovot	24	6.0	84	12.0	—	—	1
Jerusalem Mtns.	15	6.0	56	26.0	—	—	0
Lachish	27	14.0	14	23.0	—	—	5
Habsor	—	—	—	—	—	—	8
Negev	17	14.0	—	—	—	—	7
Arava	40	18.0	—	—	—	—	10
Jordan Valley settlements	4	3.0	34	16.0	—	—	2

Table 5.1.2 gives freshwater supply step functions for the districts according to their sources. The associated costs are for production and supply within the district itself. For example, it is possible to provide 150 million cubic meters (mcm) per year in the Hula district at a cost of 6 cents per cubic meter and an additional 50 mcm per year at a higher cost of 8.3 cents per cubic meter. We have assumed 662 mcm per year as the average net product of the Upper Jordan basin with all its sources north of and including the Kinneret[7] (rain plus flows from east, west, and north, minus evaporation from the lake). The cost of production at the lake is set to zero. However, the cost of supply from the lake to consumers in that district is set to 9 cents per cubic meter (see p. 84).

The final column in the table ("Local nonpotable sources") gives, for each district, the quantity of water from floods and brackish sources that is locally available for agriculture. These quantities are assumed in the model to be available at zero cost and always to be used and available.[8] They are included in agricultural demands. Although these quantities are included in the figures for 1995 agricultural consumption in Table 5.1.1 above, unless otherwise noted, they are not included in later tables that report on results of model runs.

In some cases several districts draw from the same underlying supply sources—for example, the Jordan River basin and sections of the Mountain Aquifer. This is handled by specifying that certain supply steps in the districts in question draw on a common pool, total extraction from which is limited. In the Israeli runs, the sources in the north are treated in this way. In particular, we assume that the third supply step in the Golan (25 mcm), all the Hula sources (both steps, 200 mcm), and the source at Biqaat Kinarot (the Kinneret itself, 662 mcm) are basically the same source—the Jordan River—with a total quantity of 662 mcm (see Table 5.1.2). Therefore we constrain the model to use no more than a total of 662 mcm from these sources taken together.[9]

Similarly, Beit Shean and Gilboa Harod share an aquifer, represented as the second supply step for these districts in Table 5.1.2, and the total to be pumped is constrained to 14 mcm per year.

The total possible supply of freshwater over all districts is 1583 mcm per year, which is higher than total consumption for freshwater in 1995 (1513 mcm per year[10]). Note, however, that even if total supply were lower than total demand, there would not therefore be an absolute water shortage, since the total demand cited above merely corresponds to a base point on the demand curve. The amount of water demanded depends on price, and, taking this into account, the model allocates to consumers the amounts that maximize total net social welfare. As will be seen in the results, the total amount allocated does not exceed the total available in the sources and can be even less, if conveyance constraints prevent transfers between districts or if prices higher than those used for the base points are charged to consumers.

As for recycled water, we assume the presence of recycling plants in each district with a recycling cost of 10 cents per cubic meter.[11] Furthermore, we assume that no more than 66% of freshwater used by the urban and industrial sectors can be recovered for use in recycling (as an average feasible figure).

The existing conveyance system

Israel has one conveyance system for freshwater and another for recycled water.

The following tables (Tables 5.1.3 and 5.1.4) present the conveyance systems and conveyance costs (in cents per cubic meter). Since all the links already exist, these are the operating costs and do not include capital costs. Some links connect districts to the National Carrier (marked "NC").[12] As seen in the schematic map in Figure 5.1.2, there are 33 links in the representation of the system. Each link has a specified direction of flow, given by

Table 5.1.3. Conveyance Costs for Freshwater

From	To	Cost ($¢/m^3$)
Hula	NC-Hula	0.0
Maale Hagalil	Merom Hagalil	5.4
Acco	Maale Hagalil	10.0
Biqaat Kinarot	Beit Shean	6.1
Biqaat Kinarot	Lower Galilee	25.3
Biqaat Kinarot	NC-Jezreel Valley	10.9
Jezreel Valley	Acco	5.0
Hadera	NC-Hadera	11.5
Raanana	NC-Raanana	6.5
Rehovot	NC-Rehovot	12.1
Jerusalem Mtns.	NC-Jerusalem Mtns.	12.1
Negev	Arava	25.0
NC-Golan	Golan	17.1
NC-Golan	NC-Merom Hagalil	0.0
NC-Hula	NC-Golan	0.0
NC-Merom Hagalil	Merom Hagalil	12.0
NC-Merom Hagalil	Biqaat Kinarot	0.0
NC-Jezreel Valley	Jezreel Valley	12.7
NC-Jezreel Valley	Nazareth Mtns.	15.4
NC-Jezreel Valley	NC-Hadera	0.5
NC-Hadera	Hadera	5.5
NC-Hadera	NC-Raanana	0.0
NC-Raanana	Raanana	2.4
NC-Raanana	NC-Rehovot	0.6
NC-Rehovot	Rehovot	5.2
NC-Rehovot	NC-Jerusalem Mtns.	0.0
NC-Jerusalem Mtns.	Jerusalem Mtns.	10.7
NC-Jerusalem Mtns.	NC-Lachish	0.3
NC-Lachish	Lachish	8.7
NC-Lachish	NC-Habsor	0.7
NC-Lachish	NC-Negev	0.7
NC-Habsor	Habsor	8.4
NC-Negev	Negev	8.7

Notes: It was not feasible to list (or use) all separate connections and links in detail. The links between districts listed are therefore aggregates of actual connections, and the costs are weighted values. NC = National Carrier.

Table 5.1.4. Conveyance Costs for Recycled Water

From	To	Cost ($¢/m^3$)
Acco	Jezreel Valley	6.9
Raanana	Negev	15.8
Negev	Habsor	5.1

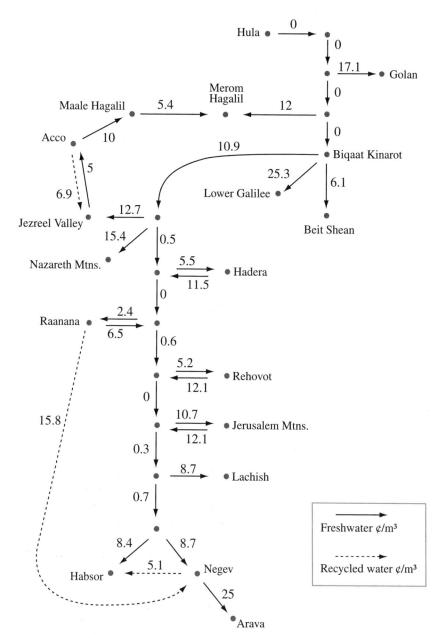

Figure 5.1.2. Conveyance Cost between Adjacent Nodes

the "from" and "to" nodes, and the associated cost refers to transfer in that direction.

The user can easily change links, capacities, and costs through the model's interface.[13]

Further constraints

To get close to real conditions, the following constraints were imposed:

- Intradistrict conveyance cost in Biqaat Kinarot is 9 cents per cubic meter. In all other districts, intradistrict conveyance is assumed to be costless.
- As indicated in Note 3 to this chapter, agricultural use of recycled and other nonpotable water in the Raanana district is limited to 5.65 mcm because of environmental restrictions (5.65 was the actual quantity used in Raanana in 1995).[14]
- We require 5 mcm per year to be supplied to Gaza from the Israeli National Carrier at Lachish, as in fact is done. We assume the conveyance cost of that supply to be 10 cents per cubic meter.
- For runs for future dates (2010 and 2020), we require 55 mcm per year to be supplied to Jordan in accord with the 1994 Israel–Jordan treaty.[15]

2. Administered Pricing versus "Competitive" Allocation

Description of runs

We begin by running the model for 1995 with fixed prices charged to consumers (as was actually the case in Israel) and then comparing the results with those obtained when prices are not fixed and different users compete for water. These two versions are called, respectively, administrative and competitive. In each case, we enter the same demand, supply, and conveyance data and let the model find the optimal allocation in terms of social welfare.

In the competitive run, given the existing infrastructure—already built and hence with its capital costs already sunk—the model allocates water to different users to maximize net private social benefits. It is as though water users in each district were charged the relevant shadow value of water. Without additional social values (such as extra value for water for agriculture), such pricing would indeed be optimal. As explained in Chapter 2, however, this does not mean that it is optimal not to charge water consumers for capital costs. Rather, it means that it is not optimal to do so in the price of water *at the margin* and thus affect water demand. Instead, capital costs should be recovered through hookup charges or some other form of fixed charge.

Of course, the Israeli government does not do that but rather sets fixed prices for water—quite high prices ($1 per cubic meter) for households and industry and relatively low ones (17 cents per cubic meter[16]) for agriculture. This policy presumably reflects the view that water for agriculture is particularly important and that nonagricultural consumers should pay high prices while agriculture is subsidized. *It is not our intention to criticize this approach.* We can, however, present estimates of the costs that such a policy entails; presumably, the benefits are judged to be worth those costs.

Table 5.2.1 gives the results of the competitive run in terms of water quantities as well as the actual (administered) quantities for 1995.[17]

As already described, in the administrative allocation, prices are set by the government. Urban and industrial users pay the price charged by the water system (Mekorot) plus all other costs (depreciation, administration, sewage disposal, etc.); the price charged to farmers does not include these additional costs and consists only of the Mekorot charge for water itself. Hence, one expects to find that in the competitive allocation, larger quantities of freshwater would be consumed in the urban and industrial sectors and less by agriculture.

This is borne out in the results. Table 5.2.1 shows that when we go from the administrative to the competitive allocation, total urban consumption rises from 541 to 631 mcm per year, and industrial consumption rises from 110 to 143 mcm per year. These changes are due to a considerable drop in the prices charged to those consumers (from $1 per cubic meter to 25 cents per cubic meter in the Jerusalem Mountains and much less than this in the other large urban areas). In agriculture, on the other hand, consumption of freshwater drops from 852 mcm per year in the administrative run to 729 mcm per year in the competitive run, while recycled and nonpotable water consumption rises from 400 to 550 mcm per year.

It is noteworthy that in going from the administrative run to the competitive run, total freshwater consumption remains the same (1503 mcm per year), with a shift of 123 mcm per year from agriculture to the other two sectors. This is more than offset by the increase of 150 mcm in agricultural use of recycled water—a shift made possible in part by the increased freshwater consumption by households and industry.

We see some exceptions to these general trends when we compare quantities by districts. In Beit Shean, for example, agricultural consumption of freshwater greatly increases relative to the administrative allocation. Indeed, in general, the model allocates more freshwater to agricultural users in the north of the country and less to southern ones in the competitive run than in the administered one. (An interesting exception is Raanana, where there is a strict limit on the extent to which recycled water can be used.)

If one thinks about this, the reason becomes obvious. Figure 5.2.1 shows the shadow values of freshwater in the administrative run (upper numbers)[18] and the competitive one (lower numbers). These shadow values are very

Table 5.2.1. Competitive Versus Administrative Water Allocations, 1995 (mcm)

District	Urban		Industry		Agriculture			
					Administrative		Competitive	
	Administrative	Competitive	Administrative	Competitive	Fresh	Recycled + nonpotable	Fresh	Recycled + nonpotable
Golan	4	4	1	1	36	9	30	9
Hula	5	6	1	1	52	28	88	26
Merom Hagalil	4	5	1	1	30	5	24	8
Maale Hagalil	8	9	1	1	7	0	0	7
Acco	48	57	13	17	42	22	16	50
Biqaat Kinarot	6	7	1	1	27	19	38	19
Beit Shean	3	4	0	0	50	73	104	71
Gilboa Harod	1	1	0	0	22	6	22	6
Lower Galilee	4	5	1	1	18	3	11	5
Jezreel Valley	8	9	2	2	20	39	11	40
Nazareth Mtns.	22	24	4	5	11	1	0	15
Hadera	29	34	24	32	114	40	70	73
Raanana	197	230	19	25	93	5	94	6
Rehovot	67	79	18	23	114	16	51	69
Jerusalem Mtns.	65	73	6	7	22	11	0	43
Lachish	12	14	12	16	56	16	37	24
Habsor	8	9	1	1	24	43	41	15
Negev	33	38	3	4	56	44	48	35
Arava	9	11	4	5	30	13	20	20
Jordan Valley settlements	11	12	1	1	28	7	24	11
Totals[a]	541	631	110	143	852	400	729	550

Total freshwater: Administrative = 1,503
Competitive = 1,503

[a] Individual columns may not add to totals because of rounding.

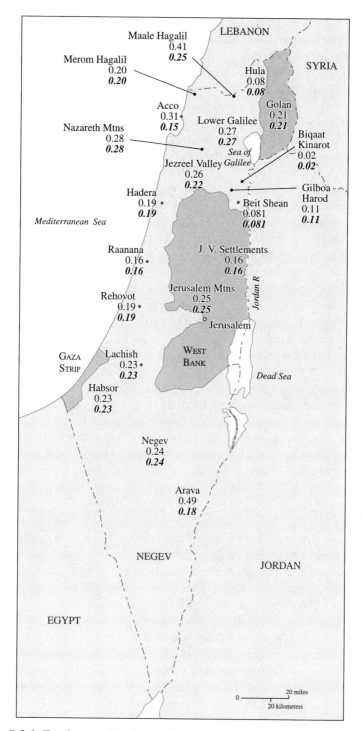

Figure 5.2.1. Freshwater Shadow Values for 1995 Administrative and Competitive Runs

similar for the two runs and largely reflect the costs of using the conveyance system efficiently. In the administrative run, because the prices charged to users are fixed, the amount of water used in each district is determined by demand considerations only. Hence, water is conveyed to southern districts (more generally, to districts farther down the conveyance system from water sources) even though the difference in shadow prices between districts is smaller than the conveyance costs involved. The policy of charging the same price to agricultural users everywhere in the country may be socially desirable, but it is not efficient in a narrow economic sense.

Those detailed results obviously depend on the data used, which may not be accurate or acceptable to the national authorities. Still, the trend of the results remains valid, as well as being quite obvious: if the only concern is economics, one should supply water to the consumers closer to the sources, where the supply costs are lower, at lower prices than to consumers farther away, where the supply costs are higher.

When we move from the administrative to the competitive run, there are also some significant changes within particular districts. In Acco, for example, urban and industrial use of freshwater increases, but agriculture uses less freshwater and more recycled water. The explanation is that the model treats fresh and recycled water as perfect substitutes in agriculture. (Incidentally, some recycled wastewater is exported from the Acco district and taken to the Jezreel Valley.) Another significant change is that the use of recycled wastewater (not including other nonpotable water) in Habsor drops from 43 mcm per year in the administrative allocation (reflecting a policy to encourage reuse) to 15 mcm per year. Recycled wastewater use in the Negev also drops (from 44 mcm per year to 35 mcm per year). These reductions result from the high conveyance costs from the Shafdan (the region near Tel Aviv) and a relatively low water-related contribution[19] for crops.

Policy analysis

As must be the case (because the optimization of net benefits is unconstrained in the competitive run and constrained in the administrative run), the calculated net benefits from water are greater in the competitive run than in the administrative one. The direct calculation of the difference is about $57 million per year (in 1995). It is important, however, to understand exactly what this figure does and does not mean.

The easiest way to think of the calculation is as follows: in the competitive run, water is charged to consumers at shadow values (plus an environmental charge to households and industry). This is the privately efficient pricing mechanism. In the administrative run, water consumers are charged fixed prices. The result is a loss of efficiency.

That deadweight loss can be depicted as follows. Consider Figure 5.2.2 (adapted from Figure 1.4.2 of Chapter 1). We saw in that chapter that the

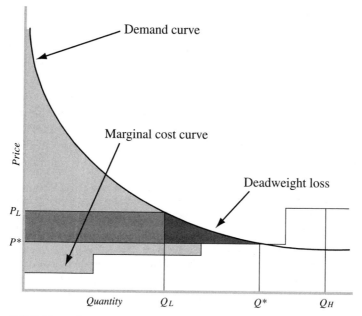

Figure 5.2.2. Transfers and Deadweight Loss from a Price above Shadow Value

net benefits from water are maximized if the quantity of water delivered to the users whose demand curve is depicted is Q^*, the quantity at which the demand curve crosses the marginal cost curve. This will also be the quantity demanded if the price charged to those users is P^*. In this simple diagram, P^* is the shadow value of water (the amount that consumers are just willing to pay for an additional cubic meter of water and the cost that would be required to produce that additional cubic meter).

Now suppose that the price charged is increased to P_L, corresponding on the demand curve to the quantity $Q_L < Q^*$. There will be the following effects:

- Consumers of water will pay more for the Q_L units they continue to consume than they did previously. That amount is represented by the area of the shaded rectangle.
- Sellers of water (possibly the government) will increase revenues by that same amount.
- Consumers will also lose because they will consume less water than before. The lost benefits from this are represented by the area of the shaded triangle.

Note that the first loss to consumers (the shaded rectangle) is exactly matched by a gain to sellers. This is simply a transfer of benefits within the country (Israel) from one group to another.

The second loss (the shaded triangle), however, is a loss that is not offset. This is the deadweight loss. It is the loss of net benefits that occurs because consumers would be willing to pay for the amount $Q^* - Q_L$ more than it would cost to produce that amount, yet that amount is not produced.

Transfers and deadweight loss also occur when there is a subsidy that affects the marginal price of water (the price charged for the last unit). Suppose that the price charged is decreased from P^*, say, to the price that would induce consumers to demand $Q_H > Q^*$. Then consumers would pay less for the first Q^* units they consumed than they did previously. On the other hand, sellers would receive less for those units; this is a transfer from sellers to consumers.

Deadweight loss in this situation arises because the additional $Q_H - Q^*$ units consumed cost more to produce than consumers would be willing to pay for them were they not subsidized—that is, they cost more to produce than the benefits they bring to consumers.

Thus, any departure from pricing at shadow values leads to a deadweight loss and also to transfers. The $57 million figure quoted above is the sum of the annual deadweight losses induced (under 1995 conditions) by Israel's system of fixed prices.

It will be useful to spell out and explain the entire accounting system involved here. Table 5.2.2 lists the items involved and gives their magnitudes. The items whose titles are in parentheses are to be subtracted when calculating total net water benefits.

The first line of the table shows the difference in buyer surplus between the two allocations. We shall discuss this in greater detail below. Note that (as remarked in footnote *a* to the table) only the difference has meaning here (as is also true for total surplus). The administrative allocation reduces buyer surplus by $370 million compared with the competitve allocation.

Not all of that difference represents a net loss to Israel as a whole, however; most of it consists of transfers. One of these, called "Freshwater scarcity rents" in the table, consists of the difference between the shadow values of freshwater and the direct costs of extraction and conveyance (see Chapters 1 and 2). If freshwater were sold in a private market at shadow values, this would be the profit of the sellers of freshwater or, equivalently, the producer surplus—the excess of payments received for freshwater less the direct costs of supply. In the administrative allocation, even though less total water is used than in the competitive allocation, there are substantial differences in the geographical pattern of water usage and in scarcity rents. (See Figure 5.2.1.)

Of course, in the administrative run, water consumers buy water not at shadow values but at prices set by the government. In effect, we account for this by treating the government as purchasing water at shadow values[20] and selling it to consumers according to the schedule of fixed prices. The difference between government inflows and outflows is labeled "Government

Table 5.2.2. Differences in Social Welfare between Administrative and Competitive Allocations, 1995 ($million)

	Administrative	Competitive	Difference (administrative − competitive)
Total buyer surplus	[a]	[a]	−370
(Government costs)	(−485)		(−485)
Freshwater scarcity rents	113	86	27
(Unaccounted-for environmental costs)	(195)		(195)
(Unaccounted-for conveyance costs)	(4)		(4)
(Net payments for constrained capacity)	(0)	(0)	(0)
Total and deadweight loss[b]	[a]	[a]	−57

[a] Because buyer and total surplus are both measured from an arbitrary origin, the absolute levels of these items have no meaning; only differences are meaningful. The distribution of changes in buyer surplus is discussed below.

[b] Figures do not add to total due to rounding.

costs." In the administrative allocation, such costs are negative because the government (net) charges consumers—especially households and industrial consumers—more than the shadow value of the water involved. Part of the loss of buyer surplus that occurs when we move from the competitive to the administrative allocation consists of an increase of $485 million in net government receipts. This is a transfer rather than a loss to Israel as a whole.

Still other items appear when consumers pay fixed prices instead of shadow values. The most important of these is "Unaccounted-for environmental costs."

As explained in Chapter 2, the model permits the recognition of an environmental cost when wastewater is produced. In all the runs we have done, that cost is set at 30 cents per cubic meter and the wastewater is assumed to be produced by households and industry. When there are no scarcity rents for recycled wastewater, and when there are no fixed prices to consumers, the optimal solution effectively passes these environmental costs through to the wastewater-producing sectors in the form of a price for freshwater that is 30 cents per cubic meter above shadow value. This reduces their buyer surpluses by the amount of the environmental costs, so those costs are included in the net benefit calculation. Under fixed prices, however, this cannot happen, so the environmental costs must appear in a separately stated line in the accounting.

The matter is a bit subtler than that, however. As also explained in Chapter 2, when there are scarcity rents for recycled water (equivalently "profits" in treating wastewater to make it safe for use in agriculture) but no fixed prices, the optimal solution effectively reduces the effluent charges to wastewater producers by allocating those scarcity rents to them. This is as it should be. When recycled wastewater has a positive scarcity rent, there is a benefit to the system from the production of such water and thus an offset to the effluent costs. (This is particularly easy to see if one thinks of the effluent costs as equal to the costs of treating wastewater for environmentally safe disposal.)

When consumers pay fixed prices, therefore, the separately stated environmental costs must be reduced by the amount of the scarcity rents in recycled wastewater. This has been done in Table 5.2.2.

There is another, equivalent way to think about this. The buyer surplus of agricultural water consumers includes benefits coming from the use of recycled wastewater. Their payments for that water are subtracted from those benefits. When prices are not fixed, those payments cover both the costs of producing the recycled wastewater and the scarcity rents of that water, the latter going to the two other consuming sectors in the form of a reduction in their effluent charges and an increase in their buyer surpluses. When prices are fixed, parallel to the treatment of freshwater, agricultural consumers' payments go to the government; the government in turn purchases the recycled wastewater at its shadow values; and the sums spent for those purchases less the costs of recycling are the scarcity rents of recycled wastewater. But those scarcity rents remain in Israel with someone—in this pro forma treatment, they end up with the producers of recycled wastewater. They must therefore be added back in to the net benefit calculation, and this is done by subtracting them from the unaccounted-for environmental costs.

The next item is not nearly so complicated as the one just explained, although the two have some features in common. The model permits the specification of nonzero intradistrict conveyance costs, and these are taken to occur in one Israeli district, Beqaat Kinarot, where they are estimated at 9 cents per cubic meter. When prices to consumers are not fixed, such costs are effectively automatically passed through to consumers in the affected district or districts, reducing their buyer surpluses. When prices are fixed, however, this does not happen, so the costs must be separately stated in the calculation of net benefits.

The last line item in Table 5.2.2, "Net payments for constrained capacity," occurs as follows. Suppose, as is not the case in either of the two runs under discussion, that the capacity of a particular pipeline (or other conveyance method) carrying water from A to B is reached. Then the shadow value of water at B will include not merely the operating costs per cubic meter of the conveyance system but also the shadow value per cubic meter of the conveyance capacity constraint. When prices are not fixed, the payments of consumers at B will include the latter shadow value (and when prices are fixed, the payments of the government will include it). But the suppliers of water at A, being paid the shadow value of their water, will not receive the payment for the scarce conveyance capacity.

Such payments do not disappear from the system, however. Whether we think of them as going to the operator of the pipeline, to the government, to the distribution company at B, or somehow to water suppliers, they go to someone. Hence they must be added back into the net benefit calculation. (This can be particularly important when the pipeline crosses an international boundary and is owned by only one of the parties involved.[21]) Similar effects must be taken into account when recycling plants are subject to a binding capacity constraint.[22]

The sum of the last column of Table 5.2.2 (subtracting items in parentheses) is the change in net benefits that are not transfers from one line to another—the deadweight loss associated with moving from the competitive to the administrative allocation.

One point that must be clearly understood is the payment of capital costs. (This will matter below but is irrelevant to deadweight loss.) As explained in Chapter 2, the shadow values generated by the WAS model do not (and should not) include capital costs. But capital costs must be paid, and presumably, this is one reason Israeli consumers are charged fixed prices.[23]

Whatever pricing scheme is used, capital costs are a charge to Israel as a whole. Charging fixed prices to collect the funds necessary to pay the capital costs is just a matter of *who* pays them, not a matter of whether they are paid. If, as suggested in Chapter 2, water consumers were charged hookup fees to recover the capital costs or were taxed directly, the same people could, in principle, pay for the capital costs who do so now under the administered-price system. But—and this is the point—there would be a net gain to Israel as a whole of $57 million per year (under 1995 conditions) were the competitive policy implemented and consumers charged shadow values.

The issue of capital cost payments appears relevant, however, when we attempt to identify the gainers and losers from the fixed-price policy. Although we can estimate the reductions in buyer surplus that occur because fixed prices rather than shadow values are used, such a comparison overlooks the fact that if fixed prices are not used to recover capital costs, such costs must be recovered from somewhere, possibly from the same consumers who pay high fixed prices. This means that the total reduction in buyer surplus given above is overstated (apart from an effect in the other direction, discussed later).

Capital costs, as such, have nothing to do with this. In the administrative run, the government receives $485 million per year. Those funds are used for some set of expenditures. If the government did not receive that money in payments for water, presumably it would still need to raise a comparable sum to finance the same expenditures. That sum would be raised in some other manner from the same entities (households, firms, and agricultural

Table 5.2.3. Changes in Buyer Surplus, Administrative Run Versus Competitive Run ($million/year)[a]

District	Households	Industry	Agriculture	Total
Golan	−3	0	2	−1
Hula	−3	−1	−6	−10
Merom Hagalil	−2	0	1	−1
Maale Hagalil	−4	−1	1	−4
Acco	−30	−8	0	−38
Biqaat Kinarot	−3	0	−2	−5
Beit Shean	−2	0	−6	−8
Gilboa Harod	0	0	0	0
Lower Galilee	−2	−1	2	−1
Jezreel Valley	−4	−1	2	−3
Nazareth	−10	−2	−1	−13
Hadera	−17	−15	3	−29
Raanana	−115	−12	−1	−128
Rehovot	−41	−12	3	−50
Jerusalem Mtns.	−31	−3	−2	−36
Lachish	−7	−8	5	−10
Habsor	−4	−1	4	−1
Negev	−19	−2	6	−15
Arava	−6	−3	0	−9
Jordan Valley settlements	−7	−1	0	−8
Total	−310	−71	11	−370

[a] Figures are rounded to nearest million dollars. Individual columns may not add to totals because of rounding.

enterprises) that consume water. Hence, the gain in buyer surplus in the competitive run would be offset (as would the loss in government revenues). How buyer surpluses of particular groups or districts end up would depend on the method used to raise the money. Note again that this is so whether or not the expenditures to be covered include water infrastructure capital costs.

Nevertheless, the comparison of buyer surpluses in the two runs is of interest, since it indicates who gains and who loses from the fixed pricing system directly. Table 5.2.3 gives that comparison, broken down by district and type of consumer.[24] The amounts given are in millions of dollars per year and are calculated as the buyer surplus for the administrative run less the buyer surplus for the competitive run. Hence, as before, a negative sign represents a loss in buyer surplus due to fixed pricing, and a positive sign represents a gain.

The results are interesting. Presumably, the use of fixed prices is intended to benefit agriculture. And this is how it appears to turn out. The agricul-

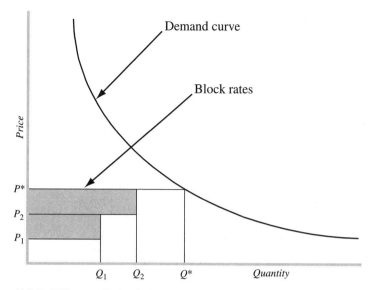

Figure 5.2.3. Effects of Block Rates on Water Consumption

tural sector is the only gainer from fixed prices, although the gain seems small ($11 million per year) compared with the losses of the other two sectors ($310 million per year for the urban sector and $71 million per year for industry). Moreover, as we shall see below, the gains for the agricultural sector are certainly overstated.

When we look at the geographic distribution of the gains and losses to agriculture, another phenomenon emerges. The major gainers and the water consumers whose consumption is greatly encouraged through the fixed-price system are the farmers in the South, starting with the Rehovot district. The farmers farther north either gain very little or lose, having their water consumption discouraged.

We have already seen this phenomenon in another way. We observed above that compared with the competitive run, the administrative allocation uses more water in the South than in the North, an effect also mirrored in the shadow values.

There is more to be said than this, however. Israeli water-pricing policy toward agriculture does not merely specify a fixed price; it specifies a price schedule of block rates for freshwater, with the earlier blocks available at lower prices than the later ones. But it is only the marginal rate ʰat matters for the decision of how much water to use. This is illustrated in

In Figure 5.2.3, water is available up to Q_1 cubic meters price of P_1 per cubic meter; an additional $(Q_2 - Q_1)$ cubic m at a price of P_2 per cubic meter; and further water at a price c meter. But the farmer's choice of how much water to use d example shown in the diagram) only on P^*. That choice is (

at which the block-rate step function crosses the demand curve and is independent of P_1 and P_2. The effect of the earlier steps is simply to subsidize the farmer relative to charging at P^* (by the amount shown in the shaded area) without changing the water-use decision at all.

To put it differently, so long as, at P^*, the farmer demands amounts in excess of Q_2, the authorities' choice of Q_1 and Q_2 and their associated prices determines only the size of the subsidy (relative to charging at P^*). It has no effect on water use.[25]

There is another, related phenomenon, however. In the late 1990s, Israel had a policy of *decreasing* block rates for the use of recycled wastewater. The first amounts of such water were priced at 14 cents per cubic meter in 1999, and later amounts at approximately 10 cents per cubic meter.

This is the parallel situation to that depicted in Figure 5.2.3, but with declining block rates. Again what matters to the water-use decision is the marginal rate. Charging a higher price for the inframarginal quantities effectively taxes rather than subsidizes agricultural use of recycled wastewater.

Unfortunately, we have been unable to estimate the size of this latter, taxing effect for 1995. We have only the rates for 1999 and no district-by-district estimate of how much recycled wastewater was bought at the higher rate. We can say, however, that the existence of this "tax" means that the differences in buyer surpluses for agriculture reported in Table 5.2.3 are overstated.[26] This could easily mean that the apparent small gains in buyer surplus for northern farmers are in fact negative.

To sum up: Israel's fixed-price policies in 1995 cost the Israeli economy $57 million per year. They benefited southern agriculture slightly at the expense of household and industrial consumers, and possibly northern farmers. The system of block rates largely subsidized or (in the case of recycled wastewater) taxed agricultural consumers without (except for the marginal rates) affecting their consumption.

One must be very careful in drawing further conclusions from all this, however. The use of the system of fixed prices instead of shadow values can be considered as having two principal intended consequences. The first of these is distributional: farmers are subsidized and other water-consuming sectors are taxed. The second effect has to do with water use. Water consumption by agriculture is encouraged, and water consumption by households and industry is discouraged.

So far as the distributional effect is concerned, there is no economic reason to accomplish that in such an inefficient manner (although there may be political reasons to do so). The same effect could be achieved by charging for water at shadow values and imposing hookup charges that are greater for households and industry than for agriculture. One could also charge at shadow values and directly tax households and industry while ~~ectly subsidizing agriculture.

But the water-use effect is not so readily dismissed. Presumably there are public benefits from agricultural water use that are not merely the private benefits received by farmers. The maintenance of green open spaces is one example; another can stem from the Zionist ideology of "return to the land." These benefits will not necessarily be achieved by all other systems of subsidizing farmers. A system of charging at shadow values but having low hookup charges for agriculture, for example, will not have the same effect on water usage as the present policy of low marginal water prices.

One can, however, make three observations. First, it would be more efficient to subsidize directly the effects that bring the added social benefits—green fields or agricultural output, for example—than to subsidize water consumption and thus distort the input choices of farmers.

Second, the system of block rates has relatively little to do with this and is, in effect, just a hidden means of subsidization (or taxation, in the case of recycled water) that does not affect water usage. Where farmers are buying freshwater on the third step of their fixed-price schedule, for example, all that matters to their water usage is the level of that step. The fact that they can purchase inframarginal water at lower rates contributes nothing toward the goal of encouraging their water consumption.

Finally, rational policy formation requires analysis of costs as well as benefits. Our results show that the fixed-price system cost Israel (in 1995) $57 million per year relative to a policy of charging at shadow values. In return (see Table 5.2.1), Israel obtained additional agricultural water consumption of 123 mcm per year of freshwater but 150 mcm per year less agricultural consumption of recycled wastewater. On the other hand, consumption by households was lowered by 90 mcm per year, and industrial consumption was lowered by 33 mcm per year. Further, the change in agricultural consumption was not positive in all districts; the geographic distribution of consumption was altered, generally in favor of the South. Policymakers should consider whether the benefits were worth the cost.

3. Projections for 2010 and 2020

Data

We now move to a consideration of the projection years, 2010 and 2020. The supply data remain the same as for 1995, and we shall experiment with changes in infrastructure. But the demand data are the principal feature that makes the projection years different from 1995. Table 5.3.1 presents demand projections (changeable by the user) on the same basis as Table 5.1.1, above. That is, the demand projections are made assuming fixed prices of $1 per cubic meter for urban and industrial sectors, and 16.5 cents per cubic meter for agriculture. The elasticities of the demand

Table 5.3.1. Demand at Fixed Prices for Projection Years 2010 and 2020[a] (mcm/year)

District	Urban		Industry		Agriculture	
	2010	2020	2010	2020	2010	2020
Golan	5.0	6.3	—	1	47.9	46.8
Hula	7.2	8.6	1	1	64.9	63.9
Merom Hagalil	6.2	7.8	1	1	31.3	31.2
Maale Hagalil	11.6	14.3	1	1	9.8	10.3
Acco	114.0	133.1	13	13	50.9	61.2
Biqaat Kinarot	7.9	9.5	—	—	41.1	40.8
Beit Shean	3.8	4.5	—	—	123.2	122.8
Gilboa Harod	1.2	1.4	—	—	21.4	21.3
Lower Galilee	6.6	8.2	1	1	11.4	9.6
Jezreel Valley	11.8	14.3	1	1	75.8	73.9
Nazareth Mtns.	32.8	39.9	4	4	5.8	7.3
Hadera	39.9	48.4	24	25	140.4	136.3
Raanana	209.8	225.8	19	20	74.0	52.8
Rehovot	78.2	88.0	18	19	103.5	101.6
Jerusalem Mtns.	79.8	89.2	6	6	33.3	35.6
Lachish	19.1	22.6	12	12	52.4	50.9
Habsor	3.5	4.8	1	1	80.8	78.7
Negev	73.8	95.7	3	4	132.8	121.4
Arava	15.6	19.5	3	4	54.2	51.3
Jordan Valley settlements	22.4	29.4	1	1	30.8	24.3
Totals[b]	750.2	871.2	109	115	1186.0	1141.7
Totals[b]	2010 = 2045 2020 = 2128					

[a] See text. Data sources for the projection years are the same as for 1995.

[b] Columns may not add to totals due to rounding.

functions are kept the same as before—that is, –0.2 for domestic, –0.33 for industry, and –0.5 for agriculture.

The demand projections are based on those made by Tahal and the Water Commission. Of course, the principal influence on the demand projections for households consists of population growth: the total population of Israel is projected to grow from 5.4 million in 1995 to 7.4 million in 2010 and 8.6 million in 2020.

With those projected data, it is possible to exemplify the use of WAS as a tool for analyzing the costs and benefits of infrastructure projects. Of course, all the examples given depend on the demand projections.

Desalination

We begin with the question of desalination plants.

The upper shadow values in Figure 5.3.1 are those obtained for 2010, assuming that hydrology is normal and desalination on the Mediterranean coast (at Acco, Hadera, Raanana, Rehovot, and Lachish) is available at 60 cents per cubic meter (including capital costs)—a cost roughly in line with current estimates. The same fixed-price policies as in the late 1990s are assumed to be in effect.[27]

Note that the upper shadow values for the coastal districts are all well below 60 cents per cubic meter. This shows that in a year of normal hydrology, desalination plants are not efficient except at a cost of around 30 cents per cubic meter or less. The same applies to imports from Turkey to the coastal districts.[28]

That changes when a drought causes a 30% drop in all naturally occurring freshwater sources. The lower shadow values in Figure 5.3.1 are all 60 cents for the coastal districts, showing that desalination plants are being used. Indeed, without some extra source of water, such as desalination or imports, there is no feasible model solution, indicating that the water demands induced by the fixed prices charged cannot be met within the range of validity of our demand curves.[29]

The requirements for desalinated (or imported) water in the Mediterranean coastal districts with a 30% reduction in all naturally occurring freshwater sources are given in Table 5.3.2.[30] They are fairly substantial, although the plant sizes suggested appear quite practical. Note that the model suggests larger plants farther up the coast than the Lachish district, where Ashkelon, often mentioned as the efficient site, is located.[31]

An important note about fixed prices: here and throughout this chapter, we assume that Israel would not adjust its fixed-price policies when there is a drought. This is not the case in practice; agricultural water quotas available at low fixed prices are reduced in drought years, so agriculture acts as a water buffer against drought. By ignoring this effect, we are making the

Table 5.3.2. Desalination (or Import) Requirements in Mediterranean Coastal Districts in 2010 with 30% Reduction in Natural Freshwater Sources and Fixed-Price Policies in Effect (mcm/year)

District	Water requirements
Acco	80
Hadera	64
Raanana	17
Rehovot	51
Lachish	29
Total	241

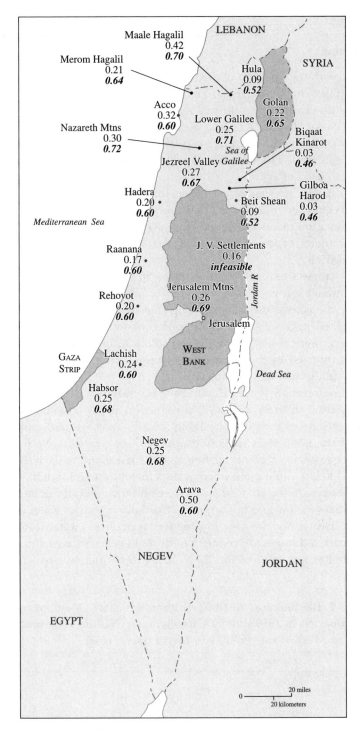

Figure 5.3.1. Shadow Values with Desalination, 2010: Normal Hydrology vs. 30% Reduction in Naturally Occurring Freshwater Sources with Fixed-Price Policies

best case for desalination and other new water sources. The actual case is not so favorable.

The situation is much less favorable to desalination on the Mediterranean Coast without the fixed-price policies, in part because drought is permitted to affect the prices paid by consumers.[32] Figure 5.3.2 gives shadow values corresponding to those in Figure 5.3.1 but without fixed-price policies. Even with a 30% reduction in all naturally occurring freshwater sources, desalination at 60 cents per cubic meter is efficient only at Acco, where the plant would require a capacity of only 39 mcm per year.[33] (It is worth remarking that even this plant would not be required were there no treaty obligation to deliver 55 mcm per year to Jordan.)

Of course, a 30% reduction in all naturally occurring freshwater sources is a severe and fairly rare event, occurring very roughly about 13% of the time.[34] We can also explore the need for desalination in lesser droughts, in particular, when there is a 20% reduction—an event that occurs roughly 20% of the time. The results are as follows:

- Without fixed-price policies, seawater desalination is not efficient on the Mediterranean coast.
- With fixed-price policies, the plant capacities required are as shown in Table 5.3.3—that is, total desalination is only about two-thirds that shown for the 30% case above (and with one fewer new plant).[35]
- There would have to be a desalination plant at Eilat producing about 17 mcm per year.

For 2020, the results are somewhat similar but more favorable to desalination, as we should expect. We find the following:

- Perhaps surprisingly, in years of normal hydrology, desalination on the Mediterranean coast is still not efficient, at costs above 34 cents per cubic meter, with or without fixed-price policies.
- In years of a 20% reduction in all naturally occurring freshwater sources, desalination becomes efficient with (but not without) fixed-price poli-

Table 5.3.3. Desalination (or Import) Requirements in Mediterranean Coastal Districts in 2010 with 20% Reduction in Natural Freshwater Sources and Fixed-Price Policies in Effect (mcm/year)

District	Water requirements
Acco	60
Hadera	25
Raanana	0
Rehovot	40
Lachish	25
Total	150

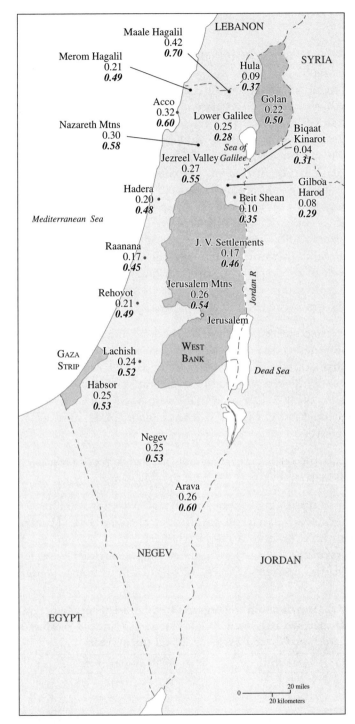

Figure 5.3.2. Shadow Values with Desalination, 2010: Normal Hydrology vs. 30% Reduction in Naturally Occurring Freshwater Sources without Fixed-Price Policies

Table 5.3.4. Desalination (or Import) Requirements in Mediterranean Coastal Districts in 2020 with 20% and 30% Reduction in Natural Freshwater Sources (mcm/year)

District	20% reduction		30% reduction	
	No fixed-price policies	Fixed-price policies	No fixed-price policies	Fixed-price policies
Acco	0	81	96	93
Hadera	0	55	0	67
Raanana	0	4	0	119
Rehovot	0	42	0	53
Lachish	0	28	6	34
Total	0	210	102	366

     cies. A total of 210 mcm per year are produced in four plants on the Mediterranean coast[36] and an additional 16 mcm per year at Eilat.

- In years of a 30% reduction in all naturally occurring freshwater sources, 102 mcm per year are produced on the Mediterranean coast without fixed-price policies and 366 mcm per year with such policies. The plant at Eilat produces 1 mcm per year without fixed-price policies and 20 mcm per year with such policies.

Table 5.3.4 summarizes the efficient amounts of desalination on the Mediterranean coast in the drought cases. Again, the size of the plants required is not unreasonable.

In the case of a 30% reduction in naturally occurring water sources, only a large plant at Acco and a very small plant in Lachish are efficient in the absence of fixed-price policies, but a very substantial amount of desalination becomes efficient in plants in other coastal districts in the presence of those policies.

Evidently, at least some of the pressure for building desalination plants comes from the expected maintenance of the fixed-price system—and (as is no surprise) from the perceived needs of agriculture, particularly in the South.

Importing water

Desalination is not the only possibility for obtaining additional water. In a peaceful world, Israel might be able to purchase such water from others—a possibility discussed more generally in Chapter 4 and explored in Chapter 8 below.

Accordingly, we now evaluate the economic feasibility of purchasing Litani River water from Lebanon in 2020.[37] We simulate this by creating a new source in the Acco district of 200 mcm per year with a cost of 33 cents per cubic meter (buying price for the water from the Lebanese plus conveyance cost of the water from the Litani River to the center of the Acco district). Then we create a new connection from Acco to the National Carrier (near Hadera) with a conveyance cost of approximately 4 cents per cubic meter and an annual capacity of 200 mcm. We assume a total investment of $573 million: $365 million for the infrastructure needed to convey the water to Acco plus $208 million for the new pipe from Acco to the National Carrier.

Obviously, in view of the results reported above, such a project would be economically inefficient if all years had normal hydrology. We already know that under such conditions, in 2010, the shadow value of water in Acco would be slightly less than 32 cents per cubic meter (see Figure 5.3.1). Hence, even were the infrastructure built, the water would not be imported. In 2020, the corresponding shadow value would be below 33 cents per cubic meter. The benefits of having water available at a fraction of a cent per cubic meter less than this would not justify the capital costs of building the required infrastructure.

In years of drought, however, the matter is far less certain, since as we have seen, in such years it would be efficient to have a desalination plant at Acco providing water at 60 cents per cubic meter. Thus, to see whether the import project would make sense in such years, we must compare the present value of the increased benefits it brings over and above the desalination scenarios with the capital cost of building the required infrastructure.

This is easily done, and we do it by taking the case of 2020, with fixed-price policies and a 30% reduction of all naturally occurring freshwater sources—of the cases examined above, the one that puts the largest strain on the system. Moreover, to make matters favorable toward the infrastructure project, we assume that imports can remain as high as 200 mcm per year even in such drought years.[38]

At least with such favorable assumptions, the import project may be worth building. The gain in net benefits would be $46 million per year. With a project life of 40 years, every one of which is assumed to have the same severe drought conditions and a discount rate of 5%, the project has a present value of approximately $800 million. Note, however, that at a discount rate of 10%, the present value of the additional net benefits from the project falls to $450 million.

We have also examined the case of a 20% reduction in naturally occurring freshwater sources, again with fixed-price policies. Here, the annual net benefits from the import project would be $43 million per year. Even with all later years assumed to have the same drought conditions, this has a present value at 5% of approximately $750 million. With a discount rate of 10%, this falls to $420 million. Evidently, the project would be worth building only with some combination of low capital costs, a long life, a low discount rate, and a very strong expectation of continued severe drought.

But a good deal of the benefits of the project would come from avoiding (at least until some later year) the building of one or more desalination plants. If such plants were to be built, then the net benefits from an import project would decrease, since once the capital costs of such plants are sunk, the cost per cubic meter of supplying the desalinated water would no longer be 60 cents per cubic meter but much lower, reflecting only operating costs. Note further that increasing population after 2020 makes it unlikely that such desalination plants can be avoided for the assumed 40-year life of the project; this would also reduce the calculated benefits.

It must be observed, however, that those results assume no international cooperation in water among Israel and its neighbors. The results might well be different with true cooperation of the water-permit type suggested in Chapter 4 and analyzed further in Chapter 8, since then imported water might benefit others in the region as well as Israel.

An important note about the value of water. Before proceeding, we pause to make an extremely important observation that emerges from the cases so far examined, each of which concerned the costs and benefits of new sources of water.

Examination of Figures 5.3.1 and 5.3.2 above shows that with normal hydrology, the shadow value of water for 2010 at Biqaat Kinarot, where water comes from the Kinneret into the Israel National Carrier, is only about 3 cents per cubic meter. In fact, for 2020, the shadow value increases by only about 1 cent.

Even with an assumed drought causing a 30% reduction in supply, that shadow value remains fairly low. For both 2010 and 2020, it varies from 31 to 39 cents per cubic meter without fixed-price policies, and is about 46 cents per cubic meter with such policies.

There are two implications here. The minor one we have already seen: in years of drought, the scarcity value of freshwater is substantially higher when fixed-price policies are in effect than when they are not.

The major implication, however, is that freshwater at the source (in this case in the Kinneret) is worth very little when there is normal hydrology. Even when there is a 30% reduction in all naturally occurring freshwater sources and desalination is efficient, the value of freshwater at its source is not very high. The waters of the Jordan River are claimed by more than one party. We have just seen that a rough calculation gives the value of 100 mcm of this water as ranging from $3 million per year to $46 million per year.[39] This is not a large number in the context of negotiations, and similar results are obtained for other disputed waters.

We shall return to this matter in Chapter 8, when we discuss regional issues and cooperation.

Expanding the conveyance line to Jerusalem

We now consider a different example of the use of WAS in the cost–benefit analysis of infrastructure.[40]

When the model is run for 2010 (normal hydrology), with no constraints on the capacities of conveyance lines,[41] the solution involves conveying 26 mcm per year from the national carrier to Jerusalem without fixed-price policies, and 36 mcm per year with fixed-price policies. For 2020, 24 mcm per year is conveyed with fixed-price policies and 27 mcm per year without.[42] This would be impossible: as of the late 1990s, the conveyance line in question had a capacity of only 17 mcm per year.

That does not make the model result ridiculous, however. What it shows is that if the conveyance line had a capacity of 40 mcm per year, it would be efficient to convey such amounts at the conveyance cost of the existing line—10.7 cents per cubic meter (exclusive of capital cost). That suggests that an expansion of the existing line is worth investigating.

Such investigation must consider the following, however:

- The efficient way to add capacity may be not to expand the existing line but to add a new one. (A detailed engineering study would be required.)
- It appears that the additional line would be at least as large as the existing one. We assume that this will raise conveyance costs of the new line (again without capital costs) to 18.3 cents per cubic meter.
- Finally, the new line would have capital costs. The question is whether the increase in net benefits it would bring would be worth those costs.

To investigate the benefits of such a project requires comparing the benefits of a conveyance system constrained to 17 mcm per year at conveyance costs of 10.7 cents per cubic meter with those obtained when the line expands to a capacity of, say, 40 mcm per year and the conveyance cost is 18.3 cents per cubic meter.

Of course, the first 17 mcm per year would be conveyed at the existing cost of 10.7 cents per cubic meter. But the amount to be conveyed will be determined by the *marginal* cost of conveyance. We must adjust the benefits to reflect the fact that 17 mcm per year is carried at a lower cost. This is easily done with a small side calculation.[43]

The results are somewhat surprising. For 2010, in the presence of fixed-price policies, the new conveyance line brings no additional net benefits. This is because the higher cost of using the new line greatly reduces the amount that it is efficient to convey below the amounts found at the lower cost with no capacity constraint. Indeed, in the presence of fixed-price policies, the additional conveyance capacity is not used at all. Hence, for the conditions of 2010, such additional capacity should not be built if fixed-price policies are to be continued.

Without such polices, there is some use of the expanded capacity and the amount of water conveyed is about 23 mcm per year. On the assumed facts, this raises benefits by approximately $5 million per year. At a 5% discount rate, this has a present value of about $90 million over a 40-year project life (almost $50 million at 10%).

The situation is different when we look at 2020. When the line is constrained to the existing capacity of 17 mcm per year, there is no feasible solution in the presence of fixed-price policies. Without those policies, there is. This reflects the fact that with fixed-price policies, the prices charged to the various sectors, especially agriculture, cause a greater demand in the Jerusalem Mountains than can be supplied with the constrained pipeline, while without such policies, the high shadow value of water in the district ($1.16 per cubic meter) is greater than the fixed-price charges, thus reducing demand.

Hence, the continuation of fixed-price policies into 2020 calls for an expansion of the line. It is interesting to note, however, that not much expansion is required. At the higher conveyance cost of $0.183 per cubic meter, only 24 mcm is conveyed.

Without fixed-price policies, 33 mcm per year is carried by the expanded system by 2020. A quantitative assessment shows that the net benefit increase from the new line is $20 million per year. If the project has a 40-year life and the discount rate is 5%, then the present value of such benefits under 2020 conditions is more than $300 million. (It is almost $200 million at a discount rate of 10%.) This means that under the assumed conditions, an additional line would be worth building if its capital costs were less than $300 million. If one expects the relevant population to increase after 2020, the present value of net benefits—and hence the capital costs that would be worth incurring—will exceed $200–300 million.

Appendix: Agricultural Submodel Results for Israel and the Use of Redundant Water Policies

In Chapter 3, when presenting the agricultural submodel, we gave results for Israeli districts in the 1990s. We now take up two other matters in the use of AGSM. The first of these is the effects of future changes in water price and availability. The second, related to our discussion of fixed-price policies, has to do with the response of agricultural production to water policies. This has application outside the Israeli context.

A.1. A Future Scenario: Hadera 2020

To demonstrate the use of AGSM as a decision support tool, the following scenario analysis is presented for the Hadera district. The scenario reflects several major changes predicted for 2020, compared with 1995:

- reduction by 60% of freshwater to 40 mcm;
- increase by 33% of recycled water to 13.2 mcm;
- increase by 10% of both brackish and surface water;

Table 5.A.1. AGSM Results for Hadera, 1995 and 2020

Selected optimal data	1995	2020
Average water price ($/m^3)	0.158	0.319
Total net income ($million)	104.008	57.375
Total water expenses ($million)	23.807	27.769
Total freshwater (mcm)	114.330	40.000
Total water demand (mcm)	150.840	87.000
Percentage of freshwater (%)	75.80	45.98
Total use of land (dunams)	298,600	238,891
Income per land unit ($/dunams)	348.32	240.17
Income per water unit ($/m^3)	0.69	0.66
Water-related contribution per water unit ($/m^3)	0.85	0.98
Unirrigated area in irrigable land (dunams)	17,400	21,500

- increase by 20% of the water-related contribution (WRC)[44] of vegetables;
- reduction of citrus area from 51,500 dunams to not more than 1,500 dunams; and
- prices of $0.40 and $0.30 per cubic meter for fresh and recycled water, respectively.

The results of the 1995 calibration run and the predicted 2020 run are shown in Table 5.A.1. The following results can be observed:

- The assumptions made about water prices lead to a doubling of average price of water in the year 2020: $0.319 compared with $0.158 per cubic meter. But income per water unit does not change by much: $0.69 compared with $0.66 per cubic meter. This means that the increase in water price will erode the effect of the assumed increase of production efficiency in agriculture (the increase in WRC). The trends in recent years agree with these results.
- Net income is reduced by 45%, very similar to the reduction in the total amount of water (42%). This reduction in net income reflects the shrinkage in Israeli agriculture during the intervening years, especially in districts like Hadera, because of urbanization (not included in the model), and the drop in net profits.
- Total cultivated land use is reduced by 20%—less than that of water— and the irrigated area is reduced by 23%. Such changes sound quite reasonable. The difference between these two values is due to the fact that the area of unirrigated crops is increased from 17,400 dunams in 1995 to 21,500 dunams in 2020.

A.2. The Response of Agricultural Production to Water Policies[45]

Background. In Israel, water is controlled by a governmental authority—the Israeli Water Commissioner (IWC). IWC's responsibilities include mining, supplying, and distributing all types of water (fresh, recycled, brackish, and surface) to all water-consuming sectors: households, industry, and agriculture. The renewable yearly amount of water in Israel is almost totally dependent on rainfall and thus is subject to irregularities, drought, and severe uncertainty regarding amounts, places, and timing. More than 95% of that total yearly amount is already being used (Kally 1997).

There are three main means by which IWC controls water: water quotas, prices, and administrative limitations on crops. The legal basis for the IWC capacity, responsibility, and power is the Israel Water Law. Regarding water pricing, that law (paragraph 112) suggests, among other procedures, that the minister in charge of IWC will decide on water prices after considering the ability of water users to pay (free translation).

As a water-consuming sector, agriculture has the following characteristics:

- It uses approximately 65% of Israel's total water amount (Kally 1997).
- It can use low-quality water.
- Agricultural demand for water is not steady but fluctuates considerably in time and space.
- The required dependability of water supply for agriculture is far lower than for the other sectors.
- Agriculture's ability to pay for water is significantly lower than that of the other sectors.
- Agriculture is very flexible in its water use and can adapt to rapid changes in water supply patterns (Amir et al. 1991, 1992).

In 1990, as the result of several consecutive drought years, a water crisis occurred in Israel. Among the responses by IWC were the following:

- a significant reduction of water quotas for agricultural use;
- an increase of water prices; and
- administrative limitations on certain crops, in particular cotton.

Those measures provoked a very noisy and bitter dispute between the authorities and the farmer organizations—which in turn gave us the incentive to analyze quantitatively the water policy measures involved.

Application. AGSM was run on agricultural production systems of 12 kibbutz settlements in the Jezreel Valley. These 12 systems, although planned and operated separately, were aggregated into one unit because water policies and water supply systems are planned and operated at a district level.

AGSM was run using 1989 data. After the calibration stage of the model, it became apparent that the actual agricultural production of the combined production system was close to the optimal production suggested by the model. The analysis regarding water policy showed four main points:

- Limiting water amounts can be a suitable measure to cope with water shortage, provided that the policy is based on an economic analysis.
- Increasing water prices, as a policy, does not necessarily lead to an increase in the productivity of water use, as is commonly thought (Eckstein and Fishelson 1994).
- Indeed, raising prices when water quotas are binding acts as an unjustified penalty, especially on efficient farmers.
- Limitations on certain crops (e.g., cotton) should be very carefully examined to avoid undesired effects on water use under certain conditions.

As we shall see, the simultaneous use of more than one method of limiting water use can lead to redundancy and unintended side effects.

The model was run in two stages. The first stage was calibration, aimed at examining and evaluating the formulation of the model and the values of the factors entered as inputs. The calibration stage was done by comparing the actual data of the year 1989 with the model results for the same input data, both per unit area and per unit water, as well as for total water and land use. The second stage, carried out after the calibration stage had shown acceptable results, was to run a series of systematic changes in the two policymaking decision variables—namely, amounts of freshwater (quotas) and water prices. The systematic runs were aimed at simulating and analyzing the trends of the response of agricultural production to these variables. For the first stage the list of crops was obviously limited to the existing ones in 1989, but for the second stage—simulation of future scenarios—more crops were added to that list. Additional runs were carried out to analyze the case in which cotton was not allowed.

Calibration—a comparison with 1989 data. The purpose of the calibration process is to examine the model's ability and dependability to reflect the real system as a decision support tool. Basically, the calibration is done by comparing the actual performance of the system in the year 1989 with results provided by the model. For this purpose the majority of the model input data used are forced to take the 1989 actual values, whereas the others are allowed to deviate from those values. The deviations from the actual experience point out possible problems in the formulation and the input data of the model. In our case, certain deviations from 1989 data were allowed in water requirements, WRC, and total areas of crops. The allowed crop area deviations were ±15% for perennial high-investment crops (such as orchards and greenhouses) and ±50% for vegetables. The industrial crops and grains areas, being low-investment and very flexible

**Table 5.A.2. Actual Data and Model Optimal Solution for Jezreel Valley,
1989**

Item	Actual 1989	Model optimum	Difference (percentage)
Irrigated winter crops (dunams)	20,569	17,738	–13.76
Industrial crops (dunams)	47,680	46,418	–2.65
Vegetables (dunams)	2,190	1,696	–22.56
Orchards (dunams)	12,333	12,333	0.00
Total irrigated land (dunams)	82,772	78,185	–5.54
Total unirrigated land (dunams)	75,884	80,471	6.04
Total land use (dunams)	158,656	158,656	0.00
Total freshwater (mcm)	21.586	21.586	0.00
Total recycled water (mcm)	10.586	10.586	0.00
Total water (mcm)	32.172	32.172	0.00
Total net income (million NIS)	29.422	29.853	1.46

crops, were free decision variables, meaning that they could have taken any
value within the system framework. The comparison between the results
of the model and the actual 1989 data in the calibration procedure is presented in Table 5.A.2.

In 1989, the prices of fresh and recycled water were 0.23 and 0.17 NIS
(new Israeli shekels) per cubic meter, respectively.

In Table 5.A.2, the following two main features can be noted:

- The differences between actual and optimal areas of all crops are within
  the range of about ±23%, even though field crops, fodder, cotton, and
  maize, which occupied 90% of the total crop area, were allowed to
  change freely. The orchard area was forced to remain unchanged.
- The difference between actual and optimized incomes (for the same
  amount of water) was 1.46%. The calculated average net income per
  cubic meter, obtained by the actual and the optimal plans, were 0.915
  and 0.928 NIS per cubic meter, respectively. Taking into account the
  inaccuracies of the input data and the limitations of a model in reflecting reality, such a difference can well be ignored. It certainly means, however, that the 12-kibbutz unified system managed to use the water quotas
  very efficiently. This was an exceptional achievement by itself because,
  in reality, all of the kibbutzim acted separately; unifying their separated
  systems was a model artifact. The fact that one can treat the district as
  an aggregate suggests that the overall water quotas for the district were
  distributed over the individual kibbutzim in a way that did not interfere
  with districtwide efficiency. As a very important part of the calibration
  procedure, the results of the calibration runs were presented to, and

Table 5.A.3. Optimal Water Amounts, Water-Related Contribution, and Productivity for Two Sets of Water Prices in Jezreel Valley (million NIS)

Water (mcm)	Net income		Water expenses		Water-related contribution		
	P_1	P_2	P_1	P_2	P_1	P_2	Difference
1	2	3	4	5	6	7	8
9.000	22.508	19.602	3.870	6.776	26.378	26.378	0
12.000	24.389	20.973	4.560	7.976	28.949	28.949	0
15.000	26.165	22.239	5.250	9.176	31.415	31.415	0
18.000	27.874	23.438	5.940	10.376	33.814	33.814	0
21.586	29.852	24.806	6.764	11.809	36.616	36.616	0

P_1: Freshwater = 0.23 NIS/m^3, recycled water = 0.17 NIS/m^3.

P_2: Freshwater = 0.40 NIS/m^3, recycled water = 0.30 NIS/m^3.

Column 1: Freshwater amounts in mcm (used as a parameter for the model runs).

Columns 2, 3: Total net income for P_1 and P_2, respectively.

Columns 4, 5: Water expenses for P_1 and P_2, respectively.

Columns 6, 7: Water-related contribution, which is net income + water cost, for P_1 and P_2 respectively.

Column 8: The difference between columns 6 and 7.

Note: The reduction in the total water amount is due to the reduction in freshwater only. The recycled water amount remains constant at 10.586 mcm. The quality of water, expressed by the ratio of the amounts of fresh and recycled water, decreases with the reduction of freshwater.

thoroughly discussed with, several managers of the agricultural production units under consideration. They approved the data, the results, and the dependability of the model.

Simulation of agricultural production—systematic runs. After the calibration stage, systematic runs of the model were carried out to simulate the optimal agricultural production response to reducing water quotas and to increasing water prices. Selected data for the Jezreel Valley, relevant to the analysis of its near-optimum agricultural production response, are presented in Table 5.A.3.

Values of income versus water amounts for the two sets of prices, taken from Table 5.A.3, are presented in Figure 5.A.1.

The upper line is for the low water prices of 0.23 for freshwater and 0.17 NIS per cubic meter for recycled water, denoted P_1 in Table 5.A.3, and the lower line is for the high prices of 0.40 and 0.30 NIS per cubic meter, denoted P_2.

Analysis. The analysis includes discussions of the three measures taken by IWC—water quotas, water prices, and land limitations on crops (cotton).

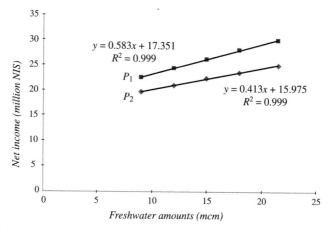

Figure 5.A.1. Net Income vs. Freshwater Amounts at Two Price Sets

WATER QUOTAS. From Table 5.A.3 and Figure 5.A.1, several conclusions can be deduced. First, income decreases as water quantities do. For P_1 the decrease of income is from 29.852 to 22.508 mill. NIS, or ΔIn = $-7.344/$ 29.852 = -24.6%, and the decrease in water amounts is from 32.170 to 19.586 mill. cubic meter, or $\Delta W = -12.584/32.170 = -39.11\%$. The decrease of income (24.6%) is lower than the decrease of water amounts (39.11%). The marginal product per water unit within the range of water amounts is ΔIn$/\Delta W$ = $-7.344/-12.584$ = 0.584 NIS per cubic meter. Such a marginal product is quite low. It is about half the value per water unit of each of the crops, the average of which (including recycled water) is 1.138 NIS per cubic meter [36.616/(21.586 + 10.586)]. The analysis shows that the moderate decrease of income, resulting from the low marginal product, could be explained by the fact that unirrigated (rain-fed) winter crops were a feasible alternative to irrigated crops. The relevant values (taken from the model) show that the net income of unirrigated winter crops is 71 NIS per dunam, and the WRC per dunam of irrigated winter crops is 168 NIS per dunam, requiring a water amount of 220 cubic meters per dunam. Net income from irrigated winter grains, for the weighted average water price of 0.210 NIS per cubic meter in 1989, was 168 – 220 * 0.210 = 121.8 NIS per dunam. The difference in net incomes between irrigated and unirrigated winter crops was 121.8 – 71 = 50.8 NIS per dunam. Although such a difference in income per dunam was quite significant, the marginal alternative contribution of water was (121.8 – 71)/220 = 0.23 NIS per cubic meter, which was quite small. Thus, growing unirrigated winter grains resulted in two outcomes: first, total income was reduced by 0.23 NIS per cubic meter; and second, the reduction in total income, 0.23 NIS per cubic meter, was relatively small compared with the average income from the crops in the optimal mix.

From the moderate reduction of income compared with the large reduction of freshwater consumption, one may conclude that for the Jezreel Valley, in which unirrigated winter crops are a real alternative to irrigated crops, the policy of administrative reduction of water quotas may be economically justified. That is, freshwater can be transferred to other districts that are willing to pay more for freshwater because they do not have the alternative of unirrigated crops.[46]

WATER PRICES. The difference between the two lines in Figure 5.A.1 expresses the difference of water expenses, D_{We}, for the two sets of water prices for every water quantity Q (from 9 to 21.586 mcm). It is calculated as follows:

$$D_{We} = Q(P_{a1} - P_{a2}) \tag{5.A.1}$$

where D_{We} is the difference between the water expenses, and Q is the water quantity.

P_{a1} and P_{a2} are the weighted average water prices of the two sets of prices, calculated thus:

$$P_{ai} = \frac{P_f Q_f + P_r Q_r}{Q_f + Q_r}, \quad i = 1,2 \tag{5.A.2}$$

where P_f, Q_f, P_r, and Q_r are water price (P) and quantity (Q) of fresh and recycled water types, respectively.

From Figure 5.A.1, it can be seen that the differences between the net incomes for the two sets of water prices are around 2 mill. NIS throughout the entire range of water quantities used. The difference in the slopes of the lines [for P_1, $(29.852 - 22.508) / (32.172 - 19.586) = 0.58$, and for P_2, $(24.806 - 19.602) / (32.172 - 19.586) = 0.41$], or $0.58 - 0.41 = 0.17$ NIS per cubic meter, is exactly the difference between the two freshwater prices $(0.40 - 0.23 = 0.17)$. This phenomenon reflects the fact that the cropping patterns were exactly the same at both freshwater prices. We discuss this in greater detail below.

Since all quotas of freshwater are used at both prices and the same amount of recycled water is used for all runs, the intercepts in Figure 5.A.1 are just the expenditures on recycled water for the two prices. This is because the regression assumes that the same fixed expenditure is made independent of water amount. The difference between the intercepts, 1.376 mill. NIS, is the difference in the recycled water prices multiplied by the (constant) amount of recycled water used (10.586 mcm).

On the other hand, the slopes are not the expenditure on freshwater per cubic meter because they involve the income being made and the costs of other factors, but with the same water amounts being used and the same cropping pattern adopted, the difference in the slopes (0.17 NIS per cubic meter) reflects the difference between the unit prices of freshwater.

From Table 5.A.3, it can be seen that the WRC for the two sets of water prices (columns 6 and 7) are the same for every freshwater amount. This is a direct result of the fact that the restrictions on the total amounts of water are binding in the solution at both sets of water prices. It also means that the optimal mix of crops does not change with the different price sets. We explain this further below.

The hypothesis that production remains almost constant under the sets of prices was further analyzed by calculating the following production ratio for all water amounts:

- water-related contribution, $\mathrm{WRC}(P_2)/\mathrm{WRC}(P_1) = 1.00$ (columns 6 and 7); and
- total water expenses ratio, $We(P_2)/We(P_1) = 1.75$ (columns 4 and 5).

The main result from the above ratios is that WRC and *We* are the same at both prices for every water amount. This means that the crop patterns are almost the same for the two price sets. Thus, raising water prices while reducing water amounts does not change the optimizing production practice but simply reduces the income of the optimizing farmers. This conclusion can clearly be verified in Table 5.A.4, which shows that for a given freshwater amount, all crop areas and water amounts are the same at both sets of prices.

The crop patterns that result from water limitation suggest the following:

- Orchard areas are stable as long as the freshwater amount is not less than 9 mcm.
- The first victims of freshwater reduction are winter crops (wheat and barley), which are changed from irrigated to unirrigated winter crops, though still located in irrigable land (from 17,738 dunams to 0). This is to be expected, because the differences in WRC between the irrigated crops and their alternatives—unirrigated crops—are significantly smaller than the original WRC values.
- The reduction in industrial crops is moderate because most of them are irrigated by recycled water, which is not subject to administrative limitations.
- The 39% reduction in water amount is the weighted average between a reduction of 58% of freshwater (67.1% of the total water amount) and a zero reduction of recycled water.
- Irrigated land is reduced by 27%, whereas the total water amount is reduced by 39%. This means that the average water requirement per area is reduced. Since the water requirements per dunam of all crops are

Table 5.A.4. Freshwater Quotas for Crops at Different Freshwater Amounts and for Two Water Price Sets

	Q = 21.586		Q = 18		Q = 15		Q = 12		Q = 9	
	P_1	P_2	P_1	P_2	P_1	P_2	P_1	P_2	P_1	P_2
Orchards (dunams)	12,333	12,333	12,333	12,333	12,333	12,333	12,333	12,333	11,625	11,625
Irrigated winter crops (dunams)	17,738	17,738	0	0	0	0	0	0	0	0
Industrial crops (dunams)	46,418	46,418	43,441	43,441	36,551	36,551	29,885	29,885	26,062	26,062
Vegetables and flowers (dunams)	1,696	1,696	1,696	1,696	1,696	1,696	1,696	1,696	575	575
Total irrigated land (dunams)	81,772	81,772	78,795	78,795	71,905	71,905	65,239	65,239	59,588	59,588
Total unirrigated land (dunams)	80,471	80,471	98,209	98,209	98,209	98,209	98,209	98,209	98,209	98,209
Total cultivated land (dunams)	158,656	158,656	155,679	155,679	148,789	148,789	142,123	142,123	136,472	136,472
Total freshwater (mcm)	21.586	21.586	18.000	18.000	15.000	15.000	12.000	12.000	9.000	9.000
Total recycled water (mcm)	10.586	10.586	10.586	10.586	10.586	10.586	10.586	10.586	10.586	10.586

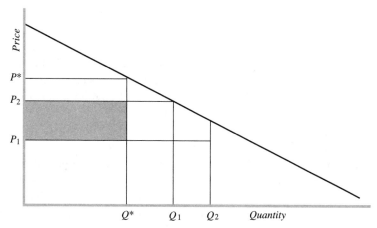

Figure 5.A.2. Schematic Demand Curve for Above-Quota Water Quantities

predetermined constants, the reduced average means a shift from more to less intensive, usually less profitable crops due to the reduction of freshwater. By checking the average WRC for the two extreme amounts of total water (Table 5.A.3), we see that the average WRC is reduced by 28%. That is, with the reduction in freshwater, crops of smaller water requirements—which also have lower WRCs—are being preferred (for example, maize is less profitable than cotton per unit area but has a higher income per water unit).

Explanation. We have found that increasing water prices does not affect the cropping pattern or the total amount of water used. Indeed, the only effect of increased water prices is to tax the income of farmers without changing their behavior. What is going on here?

The answer can be obtained by examining Figure 5.A.2, which shows the demand curve for water.[47] Here, Q^* is the amount of water allocated to the district, the "quota." P_1 and P_2 are the low and high prices, respectively, and Q_1 and Q_2 are the corresponding amounts of water that would be demanded if there were no quotas. Note that these are both greater than Q^*, indicating that the quota is a binding constraint on water use.

In this circumstance, it is clear that raising the price from P_1 to P_2 has no effect on water usage. Instead, its only effect is to increase the payments made by farmers for their allocated quota by $(P_2 - P_1)Q^*$, the area of the shaded rectangle in the figure.

Furthermore, in our runs, the demand curve generated comes from an optimizing model. Hence, given that Q^* is used at both prices, the model can be thought of as optimizing the cropping pattern with total water use fixed at Q^*. But Q^* is independent of price, hence the optimal cropping pattern must be similarly independent.

What has happened here can be described in a less detailed but more general way. Setting water allocations (quotas) and setting water prices are two ways to affect the overall use of water. Since these are two policy instruments that act on a single goal, it is likely that one of them will be effective and the other redundant. Where the water allocation is effective and water prices redundant (as a tool), the only effect of a change in water prices will be to change what farmers pay for water without changing their behavior. (In the opposite case, where prices are effective and allocations so large as to be redundant, there will be no parallel side effect.)

The income transfer effect could be controlled and water still rationed by price if the price were set to make the quantity demanded equal to the quota amount.[48] In Figure 5.A.2, that price is P^*, corresponding to the quota amount, Q^*. Since demand depends on *marginal* cost, the quota could be enforced without charging so high a price on all units; rather, the price need be charged only for units just above $Q^* - \varepsilon$, where $\varepsilon > 0$ can be as small as desired. Any lower price can be charged for the inframarginal units. Hence the income transfer effect can be as large or as small as desired.

This is a good idea. To use such a policy, the policymakers must know the price, P^*. As it happens, this can be estimated. One way of doing that is to run our model without prices but with quantity constrained to Q^* and then calculate the shadow value of the constraint. We have done this for the 1989 Jezreel Valley data for the case of freshwater (with the quantity of recycled water constrained to the actual amount of 10,586 mcm). We find values for P^* of approximately $0.59 per cubic meter for a freshwater quota of 18 mcm and $0.72 per cubic meter for freshwater quotas of both 12 mcm and 15 mcm.[49]

Note that such a policy may often require a more elaborate investigation, since there is more than one type of water. In the case investigated, we knew (from earlier runs) that a constant amount of recycled water would be used at all prices of interest; hence, we could impose that amount as a quantity constraint. In more general cases, we might have to find appropriate shadow values for both fresh and recycled water in order to use prices to ration both water types. (In even more general cases, more water types and prices would be involved.)

But if authorities wish to ration water and must decide on what prices to charge, such an exercise appears warranted and far better than arbitrarily setting both prices and quantities. The AGSM model provides one tool with which to ration sensibly. If it is so used, the question of the total amount to be paid by farmers from water can be decoupled from the instruments used to accomplish the rationing.

Limitations on irrigated crops. As already noted, the somewhat undesirable phenomena found above occur because two policy instruments act on

Table 5.A.5. Jezreel Valley: Administrative Limitation on Cotton (replacement not allowed)

Item	Area of cotton (dunams)	Total income (million NIS)	Total irrigated area (dunams)	Total recycled water use (mcm)	Total freshwater use (mcm)	Average income per total water (NIS/m³)
Cotton allowed	35,880	29.852	81,772	10.586	21.586	0.93
Cotton not allowed	0	18.861	45,892	3.146	12.586	1.20
Rate of reduction	100%	36.82%	43.88%	68.53%	41.69%	−29.03%

Note: Minus sign in the last row translates to an increase.

a single goal—the overall use of water. Where two policy instruments act on different goals, redundancy generally does not happen. Policies that affect crop choice directly can therefore be used in addition to overall policies without creating redundancies or side effects.

The third measure used by the Israeli Water Commissioner to cope with the shortage of water was in fact to limit certain crops, of which the most significant one for the Jezreel Valley was cotton. To analyze this measure, the model was run with and without cotton and with and without permitting other crops to replace cotton. Selected results of these runs are presented in Table 5.A.5.

CASE 1. COTTON REPLACEMENT NOT ALLOWED. Comparison of runs with and without cotton shows that the reductions in irrigated area and in the amount of freshwater are close to each other (43.88% and 41.69%, respectively). Total use of water was reduced by (32.172 − 15.732) / 32.172 = 51.10%. The income per water unit was increased by 29.03% (from 0.93 to 1.20 NIS per cubic meter). The reason for this increase is that the average contribution of water to the optimal mix of crops without cotton is higher than the contribution of cotton due to water. (This was the main reason for the policy of limiting cotton). However, one should take into account that cotton, which occupied 24.33% of the total cultivated area, can be also irrigated by recycled water. Because a part of the cotton in this district was irrigated by recycled water, the main outcome in this respect was that the recycled water amount was reduced by 68.53%. Therefore, the policy of taking out cotton did not in fact reduce the use of freshwater, which was the main concern of IWC, but caused a reduction only in recycled water, which the farmers were encouraged to use more intensively. Thus the policy of limiting cotton was a mistake for the Jezreel Valley. (As a matter of fact, cotton irrigated by other than freshwater types—recycled, brackish, and surface—was soon excluded from IWC's list of limited crops.)

Table 5.A.6. Jezreel Valley: Administrative Limitation on Cotton (replacement allowed)

Item	Area of cotton (dunams)	Total income (million NIS)	Total irrigated area (dunams)	Total recycled water use (mcm)	Total freshwater use (mcm)	Average income per total water (NIS/m³)
Cotton allowed	35,880	29.852	81,772	10.586	21.586	0.93
Cotton not allowed	0	27.197	81,772	10.586	21.586	0.85
Rate of reduction	100%	8.90%	0%	0%	0%	8.60%

CASE 2. COTTON REPLACEMENTS ALLOWED. Table 5.A.6 presents results, in the same format as Table 5.A.5, where cotton is not allowed but other crops (principally maize) are allowed as replacements.

Here, maize and irrigated winter crops replace cotton in the optimal mix of activities. Recycled water amounts remain at 10.586 mcm, but total income is reduced by 8.9% compared with the optimal income *with* cotton. This reduction in income, in addition to higher water prices, is a penalty inflicted on farmers, while the effect on water use, which was the trigger for the water policies, is negligible.

It seems clear that administrative limitations on cotton, when applied to the Jezreel Valley under the conditions studied, were not a good idea. The policy is an inefficient way to reduce the use of freshwater and at the same time is economically harmful to the farmers. Of course, this does not necessarily mean that such a policy may not be adequate for other conditions.

A.3. Conclusions

The effects of three kinds of water policy—water prices, administrative water quotas, and limitations on certain crops—have been analyzed using an optimizing model for agricultural production systems. From this study the following can be concluded:

• The model enables the analysis of changes in data by evaluating scenarios and water policies; thus it can be used as a decision support tool at both district and national levels.
• When water quotas are binding, raising water prices for agricultural production does not necessarily increase water productivity and efficiency, and thus may be merely a tax on the better farmers—those who practice near-optimal systems. Consequently, instead of encouraging such farmers, the combined policy of quotas and prices may contradict the basic intentions of the decisionmakers. This reflects the fact that quotas and prices are two policy instruments acting on the same goal—overall water

consumption. Their joint use is therefore likely to lead to a situation in which one of them is redundant. When that happens, there may be unintended effects.

- One way of avoiding such unintended effects would be to use our model to calculate the prices that should be charged *at the margin* to accomplish the desired rationing. The question of the total amount paid by farmers for water can then be decoupled from the instruments used to accomplish the rationing.

- The response of agricultural production systems to water limitations should be evaluated by analyzing the marginal reduction in income. In the Jezreel Valley, because of the presence of unirrigated winter crops as an alternative to irrigated winter crops, the reduction in income is small compared with the reduction in water amounts. In other words, the shadow price of freshwater is relatively small. In this case the administrative water reduction may well be justified if the water is shifted to other districts that have higher shadow values and will be prepared to pay higher prices for water.

- A policy of limiting freshwater use by applying area limitations on certain crops should take into account the type of water used by the crop under consideration. Where cotton is not irrigated by freshwater, for example, such a policy may be justified with respect to cotton when the cotton WRC is quite low. In our case, however, limiting cotton was a poor policy because it mainly reduced the usage of recycled water (rather than reducing the usage of freshwater, which was the main reason for the policy.) Indeed, we found that the policy of limiting cotton with the allowance of replacements did not change the crop pattern but penalized the farmers. The magnitude of the penalty was the difference in WRC between cotton and the replacement crops.[50]

6

Results for Palestine

This chapter begins with a brief discussion of natural features and data on water consumption, population, and available water supply, including existing infrastructure, in Palestine, followed by analysis of the current situation. Projections of future conditions are then presented, with analyses of several possible scenarios for Palestine's future.

1. Background

Palestine comprises two separated areas, the Gaza Strip and the West Bank. The Gaza Strip is bordered by Israel on the north and east and the Mediterranean on the west and shares a short border with Egypt in the south. The West Bank is the mountainous area west of the Jordan River. With the exception of the border with Jordan, the West Bank is surrounded by Israeli land.

West Bank

The West Bank is a hilly area, with elevations ranging from 400 meters below sea level in the Jordan Valley to 1,000 meters above sea level in the hills. The West Bank receives rainfall averaging about 600 mm per year, but that varies from less than 100 mm in the east and south to 700 mm in the north and west. The water supply for the West Bank relies primarily on three groundwater aquifers, a series of springs, and surface runoff. Estimates of the annual renewable quantity of water in the three aquifers vary from 590 to 690 million cubic meters (mcm). Estimates for each of the aquifers and the amount agreed upon in the Oslo II Agreement of 1995 are shown in Table 6.1.1. Of the 679 mcm, 68 mcm is brackish. The total dis-

Table 6.1.1. Estimated Recharge Rates of the West Bank Aquifers

Aquifer	Reported yield (mcm/year)	Total recharge per Article 40 of Oslo Agreement (mcm/year)
Northeastern	140–200	145
Eastern	100–130	172
Western	350–360	362
Total	590–690	679

charge of the natural springs is estimated to be between 100 and 108 mcm per year, of which 56 mcm is freshwater and the remainder is brackish.

The water quality in the West Bank is generally good, with the exception of areas with a shallow water table near either sewage discharge or agricultural activities (mostly in the Eastern Aquifer).

Surface water is intermittent in the West Bank, appearing in the form of floodwater that is difficult to capture with existing infrastructure. Approximately 70 mcm per year flows through the typically dry valley beds to the Dead Sea, and 20 mcm per year flows to the Mediterranean Sea. The Palestinians do not currently have access to Jordan River water, which is therefore not included in the available water supply for the West Bank for 1995.

Gaza Strip

The Gaza Strip depends entirely on water from the coastal aquifer that runs from the northern border of Egypt to Haifa in Israel.[1] The aquifer drains from east to west, with negligible north–south flows. Therefore the water pumped from the northern portion of the coastal aquifer in Israel has virtually no effect on the availability of water in the Gazan portion of the aquifer. However, where Israel both pumps from and returns flow to the eastern edge of the coastal aquifer adjacent to Gaza, there are potentially large impacts on water availability to Palestine.[2] Estimates of the quantities pumped from and returned to the aquifer indicate no current net effect on water availability in the Gazan coastal aquifer. Estimates of the renewable quantity of water for Gaza vary from as low as 55 mcm to as high as 100 mcm.

2. Data for 1995

Districts

For the Water Allocation System model, Palestine is divided into 13 districts, or governorates—8 in the West Bank and 5 in the Gaza Strip—that coincide as much as possible with the existing administrative units used by

Figure 6.2.1. Partial Regional Map, Palestine

the Palestinian National Authority (PNA) (see Figure 6.2.1). The 8 districts in the West Bank are Jenin, Tulkarem (including Qalqilyah), Nablus (including Tubas and Salfit), Ramallah, Jericho, Jerusalem, Bethlehem, and Hebron. The 5 districts in the Gaza Strip are Gaza North, Gaza, Deir al-Balah, Khan-Yunis, and Rafah.

Water consumption data and water demand curves

Water consumption estimates for Palestine in 1995 are based on data collected from a variety of sources, including the Palestinian Water Authority, the West Bank Water Department, the Jerusalem Water Undertaking, the Water Supply and Sewerage Authority of Bethlehem, and municipalities. In 1995 a total of 139 mcm was consumed in the West Bank, and 120 mcm in the Gaza Strip. The breakdown of consumption by sector is presented in Table 6.2.1. These data include unaccounted-for water, which is estimated to total 40% for the municipal and industrial sectors. We assume that half of this amount is from illegal connections and half is leakage due to poor infrastructure. The quantities of water consumed by Israeli settlements are not included in these estimates. All quantities of water consumed refer to freshwater only.[3]

Table 6.2.1. Water Consumption Data by Sector, 1995 (includes unaccounted-for water) (mcm)

District	Municipal and industrial	Livestock and irrigated agriculture
Jenin	3.9	5.5
Tulkarem	7.3	14.9
Nablus	8.8	25.0
Ramallah	6.7	2.1
Jericho	1.4	35.3
East Jerusalem	8.1	0.6
Bethlehem	4.3	1.4
Hebron	6.2	2.4
Gaza North	7.2	23.4
Gaza City	16.9	25.1
Deir al-Balah	7.0	16.2
Khan-Yunis	9.6	14.8
Rafah	5.4	9.2
Totals	92.8	175.9

Total consumption = 286.7

The water prices charged to consumers vary somewhat among districts because rates are set by the municipalities. However, prices in 1995 typically ranged from 20 to 40 U.S. cents per cubic meter for the domestic and industrial sectors, and we use 30 cents as an average.[4] Prices charged to agriculture are typically lower,[5] as shown in Table 6.2.2. The large differences in price in the West Bank are due to differing local access to water. In Jenin, for example, farmers can pump water directly from the aquifer, while in Tulkarem, water is available only through tanker trucks, which charge 50 cents per cubic meter.

Actual water consumption in Palestine is currently limited by the availability of supply in all West Bank districts except Jericho. The true demand curve should represent the quantity of water that would be demanded at a given price were it available. For households in these districts, this amount is estimated on the basis of per capita consumption of 36 cubic meters per year at the current average price of $0.30. This estimate is based on the per capita consumption of the unrestricted districts (assuming 7% of the combined consumption is industrial and half of the 40% unaccounted-for water is from illegal connections[6]). This quantity is also coincidentally the same as the minimum WHO 1993 guidelines for domestic consumption for small communities of 100 liters per capita per day. In addition, the price elasticity of household demand for water in Palestine is estimated at –0.6, based on the work of Mimi and Smith

Table 6.2.2. Water Price Data, 1995 ($/m³)

District	Price of water for agriculture
Jenin	0.23
Tulkarem	0.50
Nablus	0.20
Ramallah	0.50
Jericho	0.50
East Jerusalem	0.50
Bethlehem	0.26
Hebron	0.30
Gaza North	0.04
Gaza City	0.02
Deir Al-Balah	0.02
Khan-Yunis	0.03
Rafah	0.03

(2000). Estimates of demand (which typically exceeds actual consumption) are given in Table 6.2.3.

Industrial water-use data are not collected but are estimated to be a constant fraction of the total of municipal and industrial water use. Estimates range from 6% to 8%. We assume that 7% of municipal and industrial consumption is for industrial use. Table 6.2.3 gives the estimated demand for industrial water use for 1995.

The estimated agricultural demand for 1995 is taken to be the amount consumed at the prices then charged, with the recognition that the ability of farmers to respond to increased quantities of water at the prices for 1995 would have required changes in infrastructure that would not have been possible in the short run. (This issue is dealt with more explicitly in estimates of future agricultural water demand.)

For each sector and each district, the combination of price and quantity demanded define a point on the water demand curve. To obtain the complete demand curve in the relevant range, we initially assumed constant elasticities of –0.33 for industry and –0.5 for agriculture, based on available literature. In general, results are not particularly sensitive to the exact elasticity estimates used.

Supply data

The quantities available to the West Bank districts are given in Table 6.2.4. The sum of available water exceeds the total water available in the aquifers and springs because several districts have access to the same source, which acts as a common-pool resource. Constraints are placed on the supplies

Table 6.2.3. Population Data and Estimated Demand for Urban and Industrial Sectors at 1995 Prices

District	Population (1995)	Urban demand for water at 30 cents/m^3 (mcm)	Unpaid-for water (mcm)	Industrial demand for water at 30 cents/m^3 (mcm)
Jenin	182,516	5.3	1.3	0.5
Tulkarem	175,698	5.0	1.3	0.5
Nablus	289,372	8.3	2.1	0.8
Ramallah	247,990	7.1	1.8	0.7
Jericho	28,083	0.8	0.2	0.1
East Jerusalem	227,475	6.6	1.6	0.6
Bethlehem	139,925	4.0	1.0	0.4
Hebron	294,116	8.5	2.1	0.8
Gaza North	148,925	4.3	1.1	0.4
Gaza City	353,682	10.2	2.5	1.0
Deir Al-Balah	147,867	4.2	1.1	0.4
Khan-Yunis	204,030	5.8	1.5	0.6
Rafah	108,522	3.1	0.8	0.3
Total	2,548,201	73.3	18.3	6.9

Table 6.2.4. Available Water per District in the West Bank (mcm/year)

District	Available water	Oslo II quantities
Jenin	119	9.4
Tulkarem	170	22.2
Nablus	89	33.8
Ramallah	173	8.8
Jericho	57	44.4
East Jerusalem	5	—
Bethlehem	72	0.4
Hebron	62	8.8
Total	747	127.8

such that the total quantity of water pumped cannot exceed the quantity available.

The Israeli water company, Mekorot, provides limited water to the Palestinian districts in the West Bank at a price of 40 cents per cubic meter, as illustrated in Table 6.2.5. We assume that the available quantity of water to Gaza is 60 mcm and that each of the five districts in Gaza can draw on the aquifer as a common-pool resource.

The cost of water varies within districts, depending on both the cost of extracting water from a particular source (e.g., water pumped from a cer-

Table 6.2.5. Mekorot Supply to the West Bank (mcm/year)

District	Water supply
Jenin	3.1
Tulkarem	0
Nablus	3.0
Ramallah	6.5
Jericho	4.5
East Jerusalem	5.0
Bethlehem	8.4
Hebron	0

Source: http://www.arij.org/pub/water/part1.htm.

Table 6.2.6. Supply data, 1995

	Step 1		Step 2		Step 3	
District	Quantity (mcm/year)	Cost (¢/m³)	Quantity (mcm/year)	Cost (¢/m³)	Quantity (mcm/year)	Cost (¢/m³)
Jenin	67.0	23	19.8	23	32.4	23
Tulkarem	140.4	21	14.5	21	15.3	21
Nablus	18.5	20	8.6	20	7.9	20
Ramallah	118.8	28	45.6	28	8.6	28
Jericho	8.0	24	14.9	24	34.4	24
East Jerusalem	3.6	10	1.2	10		
Bethlehem	21.6	26	43.0	26	7.2	26
Hebron	19.0	28	43.2	28		
Gaza North	30.6	3.3				
Gaza City	45.1	1.4				
Deir al-Balah	23.2	1.8				
Khan-Yunis	24.5	3.1				
Rafah	14.6	3.2				

tain depth in an aquifer versus spring water, which flows to the surface naturally) and the costs of delivering the water to demands within the district. Table 6.2.6 defines the freshwater supply step functions for the districts according to their sources.[7] Note that Israel delivered 5 mcm of water to Gaza in 1995 from the Israeli national carrier at a price of 40 cents per cubic meter.

Existing conveyance

Palestine currently has no interdistrict conveyance. Intradistrict conveyance for industrial and domestic water use exists but suffers from serious problems of unaccounted-for water, at an approximate rate of 40% for these two sectors. We assume that half the unaccounted-for water is attributable to leakage, and the remaining half to unpaid-for water—that is, illegal connections.

3. Model Results for 1995

Base case results

The baseline run for 1995 allowed overabstraction—that is, overpumping—in the Gaza Strip, constrained pumping in the West Bank as imposed by Israel that year, and the additional supplies provided to the West Bank from Mekorot, as shown in Table 6.2.5. The quantities consumed in each of the districts is compared with reported consumption in Table 6.3.1, and the range of shadow values based on these assumptions is illustrated in Figure 6.3.1. Note that these quantities include the unpaid-for water, which imposes a cost to society on the order of $10 million. The quantities shown in the table below are relatively close, particularly for agriculture. The quantities for all sectors appear lower in the model, with the exception of Jericho, but with the 20% leakage accounted for, the values are representative of 1995. Jericho may, in fact, have lower levels of leakage than other West Bank districts.

Table 6.3.1. Comparison of Model and Actual Consumption, 1995 (mcm)

District	Domestic and industrial consumption		Agricultural consumption	
	Actual	Model	Actual	Model
Jenin	4	4	6	4
Tulkarem	7	4	15	15
Nablus	9	8	25	22
Ramallah	7	6	2	2
Jericho	1	1	35	35
East Jerusalem	8	6	1	1
Bethlehem	4	4	1	1
Hebron	6	7	2	2
Total, West Bank	47	40	87	82
Gaza North	7	5	23	20
Gaza City	17	13	25	27
Deir Al-Balah	7	5	16	14
Khan-Yunis	10	7	15	13
Rafah	5	4	9	8
Total, Gaza	46	34	89	82

Figure 6.3.1. Shadow Values for 1995 Base Conditions before Leakage

Leakage obviously plays an important role in water management in terms of the amount of water that ultimately reaches a consumer. More subtle is the effect leakage can have on decisionmaking for infrastructure, particularly interdistrict pipelines. If the shadow values between any two districts differ by more than the marginal cost of conveying water between the districts, it may be cost-effective to build a conveyance line between those two districts. The decision to construct the pipeline will be positive if the benefits obtained from the conveyed water outweigh the capital cost of construction. However, it is critical that this comparison of shadow values between districts happen with values *before* leakage, since the greater the intradistrict leakage in a district, the lower will be the shadow value of water conveyed there. A comparison of shadow values after leakage, which are necessarily higher, would therefore be misleading because the actual gain in social welfare is that represented by the before-leakage shadow values.

The difference that leakage makes in shadow values is illustrated in Figures 6.3.1 and 6.3.2. All later shadow values in this chapter are before leakage.

In Figure 6.3.1, the before-leakage shadow values in Gaza are consistent with the low cost of pumping in the shallow coastal aquifer. The Mekorot supply, which serves all West Bank districts except Tulkarem and Hebron,

Figure 6.3.2. Shadow Values for 1995 Base Conditions after Leakage

is used in the model in Jericho and Jerusalem, as indicated by the 40 cents per cubic meter shadow value. The Mekorot supplies are not used in the model run in Jenin, Ramallah, Bethlehem, and Nablus, however, since the shadow values are below the Mekorot price of 40 cents per cubic meter. The supplies were in fact used in these districts, however. There are several possible explanations for the discrepancy between model results and reality. One is that our assumption of zero intradistrict conveyance costs is inaccurate: if these costs are, for example, 13 and 20 cents per cubic meter for Bethlehem and Nablus, respectively, Mekorot water would be used in the model solution. Another potential explanation is that the estimate of leakage at 20% for these two districts may be too low.

In the *after*-leakage shadow values of Figure 6.3.2, the shadow value in Bethlehem has risen from 26 cents to 32.5 cents per cubic meter and in Nablus from 20 cents to 25 cents per cubic meter. The level of leakage required to reach the 40-cent Mekorot price can be computed by dividing the before-leakage shadow value by the Mekorot price. In the case of Bethlehem this rate is 35%, which is still below the reported total unreported losses to the system of 40%.

Recall that the base scenario includes the Mekorot supplies, and there is no constraint on the coastal aquifer in Gaza. As we begin to look toward the

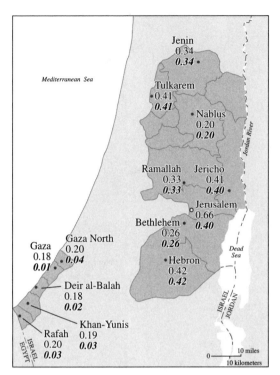

Figure 6.3.3. Comparison of Shadow Values for 1995 Base Conditions (lower values) and Constraints on Gaza Coastal Aquifer with Loss of Mekorot Supply in West Bank (upper values)

future, it is important to address both issues. The consequences of the over-pumping in Gaza are serious. Saltwater intrusion is already occurring in portions of the aquifer, leaving the water unfit for drinking and damaging to crops. An alternative scenario for 1995 imposes limits on the pumping in the coastal aquifer to sustainable levels of 60 mcm per year and eliminates supply from Mekorot to the West Bank. The 5 mcm per year supply to Gaza from Israel is continued, as this is clearly established by agreement. Not surprisingly, imposition of these constraints leads to higher shadow values in Jerusalem and Jericho in the West Bank (with Jerusalem affected by the loss of supply from Mekorot, and Jericho, which shares water with Nablus and Jenin through a common-pool aquifer, affected indirectly through the loss of Mekorot supplies to those two districts) and throughout Gaza, as illustrated in Figure 6.3.3. The shadow value in Jerusalem is relatively high at 66 cents per cubic meter, indicating an approaching crisis if nothing is done, but generally the shadow values in the coastal (Gazan) districts are not sufficiently high to warrant desalination, assumed here to be 60 cents per cubic meter. One other observation is that there are relatively large differences in shadow values between districts—notably Bethlehem and

Jerusalem, with a difference of 40 cents per cubic meter—indicating the possible benefits in conveying water from Bethlehem to Jerusalem. Conveyance will be explored in more detail in the following section.

4. Projections for 2010 and 2020

Water demand curves

The Palestinian economy is currently characterized by a small industrial sector that represents 6–12% of the total output, a service sector that represents 45–55%, an agricultural sector that represents 20–35%, and a construction sector that represents 18% of total output. The total output has varied considerably from year to year since 1980. The state of the economy will ultimately determine the demand for water for each of these sectors over time. Although Palestine has seen real declines in both total and per capita gross domestic product recently, we make the assumption that this trend will reverse and GDP will increase by 5–7% per year.[8] The assumptions relevant to each sector for the projections are described below.[9]

Urban and domestic sector

Estimates of water demand for 2010 and 2020 depend largely on the rate of population growth in Palestine, as well as issues of migration and immigration, particularly with the possibility of large numbers of returnees. Further, it is well established that consumption of water increases with increasing income. Hence, another major factor in urban demand will be the effect of changing incomes of Palestinians. There is a high degree of uncertainty about the patterns of change in population and gross domestic product for Palestine at this time. Of the various official scenarios of population growth, we use a "middle" scenario as our baseline data and explore the implications of different population estimates in our analyses later in this chapter.

Population projections are based on the 1995 Palestinian Central Bureau of Statistics estimates of population and growth rates. The growth rates for these estimates are derived from past surveys of population growth rates in the West Bank between 1967 and 1992 (PCBS 1994). These surveys indicate that such rates generally fluctuated between 1% and 4%. The large variation is due to changing political circumstances during that time. The assumed rate of population growth from 1996 to 2000 is 3%. For 2000 to 2020, the population growth rates is assumed to be 2.5%.

Returning refugees are assumed to total 500,000, with 400,000 returning to the West Bank and 100,000 returning to the Gaza Strip by 2010, and are assumed to have the same growth rate as the population in Pales-

Table 6.4.1. Population Projections for Middle Scenario for Palestine

District	2010	2020
Jenin	325,800	417,100
Tulkarem	313,600	401,500
Nablus	516,600	661,300
Ramallah	442,600	566,700
Jericho	50,200	64,200
East Jerusalem	406,000	519,800
Bethlehem	249,900	319,800
Hebron	525,100	672,100
Gaza North	240,700	308,100
Gaza City	571,900	732,100
Deir al-Balah	239,000	305,900
Khan-Yunis	329,900	422,300
Rafah	175,500	224,700
Total	4,386,800	5,615,600

tine. Returnees are allocated based on the current proportions of population in the districts. The population projections by district are shown in Table 6.4.1.

These projections are inherently uncertain but can readily be adjusted through the WAS interface so that the effects of different rates of population growth can be explored.

The projected increase in GDP discussed above is assumed to translate to an overall increase in income for Palestinians. Combining the population growth rates and the GDP growth rates defines the corresponding per capita growth rates. Although real GDP growth rates are relatively high, they translate into reasonable per capita growth rates because the population growth rates are also high.

The effect of increases in incomes on the per capita demand for water is represented by the income elasticity for household water demand.[10] The per capita demand estimates are generated by raising the estimated ratio of each year's income to the base year's income growth to the power given by the income elasticity and multiplying the result by base year consumption. Starting with the estimated initial demand in 1995, water demand (at the same price) can be estimated using the equation:

$$\text{PCWD}_t = \text{PCWD}_{1995} \, (1 + u)^{e(t-1995)} \tag{6.4.1}$$

where

u = per capita income growth rate,
e = income elasticity coefficient,

Table 6.4.2. Projected Water Demand for Palestine at 1995 Prices

Period	GDP (annual percentage growth)	Population (annual percentage growth)	Per capita GDP (annual percentage growth)	Income elasticity	Water consumption (annual percentage growth)	Projected annual per capita demand in last year of period (m^3)
2000–2010	7.20	2.5	4.7	0.50	2.4	46.9
2010–2020	6.30	2.5	3.8	0.25	0.9	51.7

Note: This study was done in the late 1990s, and the latest population survey took place in 1997; therefore the population in 2000 remains an estimate.

Table 6.4.3. Projected Urban Demand for Palestine at 1995 Prices (mcm)

District	2010	2020
Jenin	15.29	21.54
Tulkarem	14.72	20.74
Nablus	24.24	34.16
Ramallah	20.77	29.27
Jericho	2.36	3.32
East Jerusalem	19.05	26.85
Bethlehem	11.73	16.52
Hebron	24.64	34.71
Gaza North	11.30	15.91
Gaza City	26.84	37.81
Deir al-Balah	11.22	15.80
Khan-Yunis	15.48	21.81
Rafah	8.24	11.61
Total	205.87	290.05

Note: The issue of unpaid-for water is assumed to have been eliminated in all future runs.

t = time 2000, 2010, and 2020, and
$PCWD_{1995}$ = per capita water demand in 1995.

The resulting calculations for Palestine are shown in Table 6.4.2.[11] Projections of GDP and populations are from the Ministry of Planning and the Palestinian Central Bureau of Statistics, respectively. Note that the income elasticity is assumed to be lower in the first 10-year period than in the second because of a declining response with increasing income. The population projections were multiplied by the corresponding per capita water demand to arrive at total household demand at 1995 prices, shown in Table 6.4.3.

Table 6.4.4. Projected Industrial Demand for Palestine, Middle Scenario, at 1995 Prices (mcm)

District	2010	2020
Jenin	1.9	4.1
Tulkarem	1.8	4.0
Nablus	3.0	6.5
Ramallah	2.6	5.6
Jericho	3.0	0.6
East Jerusalem	2.4	5.1
Bethlehem	1.5	3.2
Hebron	3.1	6.6
Gaza North	1.1	1.5
Gaza City	2.7	3.5
Deir Al-Balah	1.1	1.5
Khan-Yunis	1.6	2.0
Rafah	0.8	1.1
Total	27.0	45.0

Industrial sector

Absent detailed economic and industrial development plans for Palestine, it is difficult to base estimates of future commercial and industrial demand on economic projections. As a result, projected industrial and commercial water demand is calculated as a percentage of the total municipal and industrial water demand. Future commercial and industrial water demand quantities have been estimated to be 7%, 10%, and 12% of the total municipal and industrial demand for the base scenario for the years 2000, 2010, and 2020, respectively. The Palestinian Water Authority has estimated slow industrial growth between 1996 and 2000, rapid growth between 2000 and 2020, and moderate but stable growth thereafter. This scheme is designed to sustain the growth of existing industries and facilitate phased development implementation of possible industrial zones. A summary of projections of future commercial and industrial water demands is presented in Table 6.4.4.

Agricultural sector

The amount of land available for agriculture, the portion of land irrigated, and the choice of crops will all influence the quantity of water demanded by the agricultural sector. A change in population, whether an increase or a decrease, will affect the availability of agricultural land and therefore the quantities of water demanded. Estimating the net effect of all of these influ-

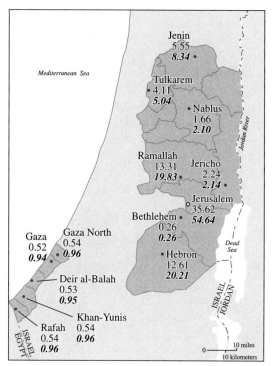

Figure 6.4.1. Shadow Values for 2010 (upper) and 2020 (lower) Base Conditions before Leakage

ences is very difficult. We explore the effects of differing assumptions about agricultural demand in the following section.

Baseline runs for 2010 and 2020

The baseline runs for future scenarios initially assume that Mekorot deliveries to the West Bank are discontinued,[12] and water is pumped sustainably in Gaza. The first scenario for 2010 assumes no change from the constrained scenario for 1995 (Figure 6.3.3, upper values), other than the increase in demands as discussed in the previous section. The shadow values in this scenario indicate a crisis in most West Bank districts, as shown in the upper numbers in Figure 6.4.1. Conditions in 2020, illustrated in the lower numbers in the same figure, are even worse. If such crises are to be avoided, either demands need to decrease or supplies increase. An obvious option to consider for new infrastructure based on this baseline scenario would be a pipeline from Bethlehem, which apparently has sufficient local supply, to Jerusalem, which has untenably low local supply. The next several sections look at several ways to increase supply—treatment plants, conveyance systems, and desalination—within the current constraints on water availability.

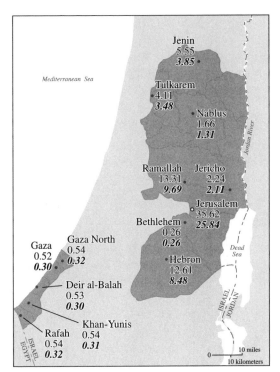

Figure 6.4.2. Comparison of Shadow Values for 2010 Base Conditions (upper) and Addition of Recycling Plants (lower)

Treatment facilities

The introduction of recycled water could potentially have a large impact on the overall availability of water in Palestine because it effectively allows for a second use of part of the water used by urban and industrial users. It is possible to use WAS to evaluate the effect of constructing a treatment facility in each of Palestine's districts. This is done in the scenario shown in the lower numbers in Figure 6.4.2. It is assumed in this scenario that 66% of the water consumed by urban and industrial users is available for recycling in the agricultural sector. The incremental cost of providing the additional treatment necessary for reuse is 10 cents per cubic meter. No capacity constraints are placed on the plants, to help determine the optimal capacities for each plant. The results show that shadow values are consistently lower than without recycling (Bethlehem is an exception) but still unacceptably high in the West Bank. Conveyance is essential. Desalination would not be needed in Gaza at an assumed cost of desalination of 60 cents per cubic meter in 2010.

There remains the question of whether the benefits of constructing these plants outweigh the costs. The increased benefits to Palestine for this series of treatment facilities are $242 million annually in 2010 and $448 million annually in 2020. Applying a discount rate of 5% and assuming a 25-year life of the plants, the present value of benefits is about $2.5 billion in 2010 and $4.5 billion in 2020, which certainly suggests further exploration of the construction and operation and maintenance costs of these facilities. The capacities used for each district in the model can be used as a guide for appropriately sizing the plants.[13]

Conveyance systems

The following tables present potential conveyance systems and conveyance costs, which have been developed based on several assumptions:

- Pipe friction is 0.006.
- Velocity varies from 1.5 to 2 meters per second.
- Operation is 20 hours per day.
- Power cost is 6 cents per kWh.
- 0.004 kWh is needed to pump 1 cubic meter of water over 1 meter of vertical distance.
- Engineering and contingencies are 15% and 25% of the investment cost, respectively.

Tables 6.4.5 and 6.4.6 represent operation and maintenance costs, based on the distances and head differences (vertical distances) between districts among the districts in the West Bank and Gaza, respectively. The estimated cost for conveyance from Hebron to the Gaza Strip is 13.4 cents per cubic meter, and from the Gaza Strip to Hebron is 49.2 cents per cubic meter.

Similarly, the capital investment required to construct each pipeline is given in Tables 6.4.7 and 6.4.8 for pipelines within the West Bank and Gaza, respectively. The capital investment required for a pipeline between Hebron and Gaza is $150 million. In developing these estimates, pipeline capacity is assumed to be 50 mcm per year, with the exception of the pipeline between Hebron and Gaza, which has an assumed capacity of 200 mcm per year. Both the head difference and the horizontal distance between districts have been used to establish these estimates. Additional costs of building pumping stations have not been included in these estimates because they are assumed to be relatively small.

The potential value of constructing conveyance systems is explored in a scenario in which Palestinian national carriers operate throughout the West Bank and Gaza, and a pipeline is built between the two areas. Two sets of pipelines are "constructed" in the model, allowing water to be conveyed in either direction between any two districts. This allows the exploration of optimal conveyance. Figure 6.4.3 shows the shadow values with conveyance in place (upper numbers), compared with the baseline scenario (lower

Table 6.4.5. Operation and Maintenance Costs for Conveyance within the West Bank (¢/m³)

	To							
From	Jenin	Tulkarem	Nablus	Ramallah	Jericho	East Jerusalem	Bethlehem	Hebron
Jenin		11.8						
Tulkarem	11.8		14.9					
Nablus		5.3		25.2				
Ramallah			25.2		7.2	12.5		
Jericho				44.4		36.7		
East Jerusalem				7.7	4.3		5.0	
Bethlehem						5.0		15.8
Hebron							11.0	

Table 6.4.6. Operation and Maintenance Costs for Conveyance within the Gaza Strip (¢/m³)

	To				
From	Gaza North	Gaza City	Deir Al-Balah	Khan-Yunis	Rafah
Gaza North		0.9			
Gaza City	1.9		6.2		
Deir al-Balah		6.2		4.4	
Khan-Yunis			2.5		2.2
Rafah				2.6	

Table 6.4.7. Required Capital Investment for Conveyance within the West Bank ($million)

	To							
From	Jenin	Tulkarem	Nablus	Ramallah	Jericho	East Jerusalem	Bethlehem	Hebron
Jenin		56						
Tulkarem	56		48					
Nablus		48		120				
Ramallah			120		120	48		
Jericho				120		72		
East Jerusalem				48	72		24	
Bethlehem						24		64
Hebron							64	

Table 6.4.8. Required Capital Investment for Conveyance within the Gaza Strip ($million)

	To				
From	*Gaza North*	*Gaza City*	*Deir Al-Balah*	*Khan-Yunis*	*Rafah*
Gaza North		6			
Gaza City	6		29		
Deir Al-Balah		29		16	
Khan-Yunis			16		11
Rafah				11	

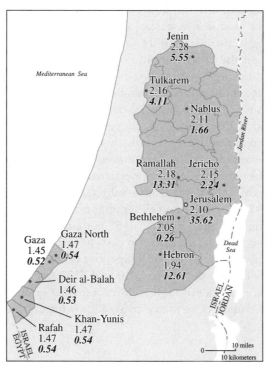

Figure 6.4.3. Comparison of Shadow Values for 2010 Base Conditions (lower values) and Addition of Conveyance Systems (upper values)

Table 6.4.9. Pipelines and Capacity Used in 2010 Scenario in Palestine

Origin district	Destination district	Quantity transported (mcm)
Tulkarem	Jenin	5.3
Nablus	Tulkarem	13.8
Jerusalem	Ramallah	13.1
Jerusalem	Jericho	9.5
Bethlehem	Jerusalem	37.9
Hebron	Bethlehem	12.0
Gaza North	Hebron	25.9

numbers). As with the introduction of recycling plants, prices generally fall, except in Nablus and Bethlehem in the West Bank, where the shadow values are low, and the five districts of Gaza, which now are supplying districts with previously high shadow values.[14] Shadow values are now more consistent throughout Palestine, with differences attributable to transport costs.

The pipelines used in this scenario are shown in Table 6.4.9. Note that 26 mcm is being transported from Gaza to Hebron—a relatively expensive option, but given the crisis without conveyance, this is appropriate. The capital cost of these pipelines can be computed from Tables 6.4.7 and 6.4.8, for a total of $462 million. The annual benefit from these pipelines is $245 million, and thus they would fully be paid for in two years. The need for conveyance under these circumstances could not be more clear. The need in 2020 is even more transparent, with an annual benefit of $375 million.

An obvious scenario to consider is combining pipelines and recycling systems. This scenario, with the upper shadow values, is compared with pipelines alone, with the lower shadow values, in Figure 6.4.4. Although this combination is clearly better than either option alone, shadow values remain unacceptably high, ranging from $1 to $2 per cubic meter in 2010, and $2 to $3 in 2020—well above desalination costs in Gaza. The incremental gain in annual benefits when going from pipelines alone to both pipelines and recycling plants is $104 million in 2010 and $170 million in 2020. Again, these benefits are sufficiently high to warrant further evaluation of the capital and operation and maintenance costs of these infrastructure plans.

Desalination

Desalination along the Mediterranean coast in the Gaza Strip is a potential new water source for the future. The previous series of scenarios considering construction of pipelines and treament facilities show that shadow values are much higher than the cost of desalination (see Figure 6.4.4), assumed here to be 60 cents per cubic meter, inclusive of capital costs. Desalination in the Gaza Strip is added to the scenario with pipelines and recycling and is compared with the same scenario without desalination in

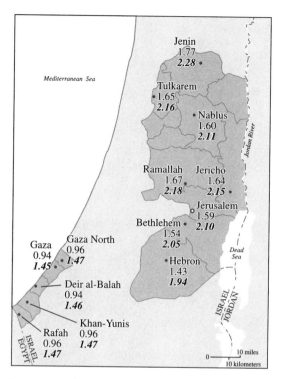

Figure 6.4.4. Comparison of Conveyance Systems Alone (lower values) with Addition of Recycling Plants (upper values), 2010

Figure 6.4.5. Shadow values in Gaza drop approximately 40 cents per cubic meter, from $1 to the cost of desalination, 60 cents.

There is an interesting phenomenon in Gaza. The northernmost and southernmost districts have desalination plants in place, but the interior districts do not have desalination, and there is no transport of water between these districts via pipelines. How did the shadow values fall from almost $1 to roughly 60 cents per cubic meter in these districts? Recall that there is a common-pool constraint on the coastal aquifer in Gaza that stretches across the five districts. The desalination plants in the north and south essentially take pressure off the aquifer, allowing the interior districts to pump additional water while still complying with the total constraint of 60 mcm.

The situation in 2020 is similar to that of 2010, as shown in Figure 6.4.6. Shadow values are slightly higher in the West Bank, but otherwise identical to 2010, as the source in this case—desalination—is unlimited at a fixed cost of 60 cents per cubic meter. Although desalination clearly improves the situation and is cost-effective, shadow values are high enough to warrant concern about the viability of such a situation. Additional sources of supply are therefore explored in the following section.

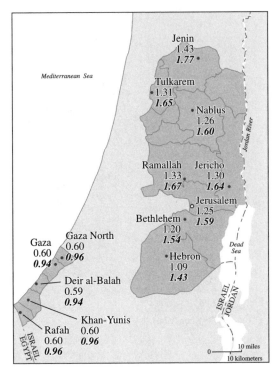

Figure 6.4.5. Comparison of Conveyance–recycling Scenario with (upper values) and without (lower values) Desalination, 2010

Of course, it is not regionally efficient to have water pumped uphill from Gaza to the southern West Bank. But without cooperation with Israel or additional sources of water, this will make sense for Palestine alone (provided, of course, that the right-of-way can be reliably obtained).

Changing availability of water from the Mountain Aquifer

The possibility of future treaties granting more water to the Palestinians is explored in a scenario by doubling the amount of water available from the Mountain Aquifer in 2010 and 2020. The effect on shadow values relative to the scenario with full infrastructure available (recycling, pipelines, and desalination) is shown in Figure 6.4.7 for 2010. Shadow values drop dramatically throughout Palestine. Desalination is no longer needed, since it is no longer efficient to transport water from the Gaza Strip to the West Bank.[15] Shadow values are relatively reasonable in the West Bank, although still too high for standard irrigated agriculture. The shadow values in 2020 (upper) are compared with those in 2010 (lower) in Figure 6.4.8. The 2020 shadow values are greater throughout Palestine, and a small amount

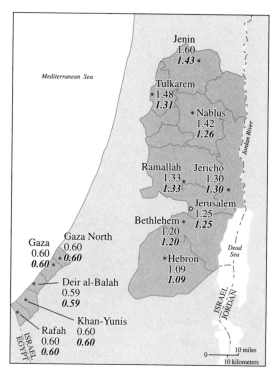

Figure 6.4.6. Comparison of Conveyance–Recycling–Desalination Scenario, 2010 (lower values) and 2020 (upper values)

of desalination is required in Gaza. However, there is still no transport of water between Gaza and the West Bank.

Changing availability of water from Lake Tiberias[16]

The other major source of water accessible to the West Bank is the Jordan River. The headwaters of the Jordan are in Syria and Lebanon, with three major tributaries: the Banyas, the Dan, and the Hasbani Rivers. These three tributaries join in the Hula Valley and form the Upper Jordan, which flows into Lake Tiberias. Below the lake, the Jordan is joined by the Yarmouk. How much of the water from this system will be available to the Palestinians will depend on the outcome of negotiations on the Jordan with Palestine's neighbors.

One possible scenario for the use of the Jordan River involves the construction of a pipeline from Biqaat Kinerot to Jenin, which would provide 150 mcm annually at a cost of about 12 cents per cubic meter. The quality of the water is assumed to be sufficiently high that desalination or other additional treatment is not required (although this could be readily handled through a cost added to the 12 cents per cubic meter). This additional

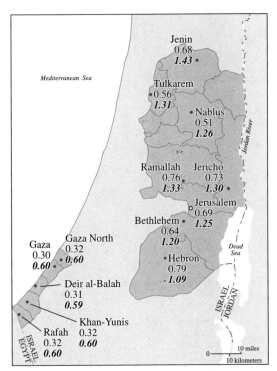

Figure 6.4.7. Comparison of Full-Infrastructure Scenario with (upper values) and without (lower values) Double the Quantity from the Mountain Aquifer, 2010

supply reduces shadow values throughout the West Bank, as illustrated in Figure 6.4.9. The most dramatic change occurs in the northern West Bank. Of the 140 mcm available, however, only 90 mcm is used. In 2020, still only 106 mcm is used, indicating a possibility of trade with Israel if Palestine in fact owned this quantity of water. Such possibilities are explored further in Chapter 8, on cooperation.

Additional scenarios

WAS can explore different assumptions about the future, which is inherently uncertain. Three assumptions for the scenarios above concern population, the amount of irrigated agriculture, and droughts. Each of these is explored further below, using the scenario with full infrastructure and doubling the quantity of water from the Mountain Aquifer as the baseline.

Increased population

The official high population estimate is approximately 30% higher than the middle estimate used in the previous set of scenarios. This increase

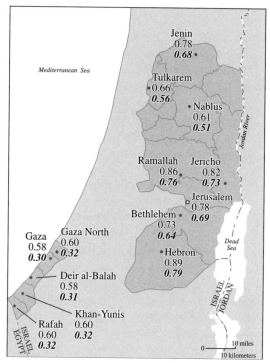

Figure 6.4.8. Comparison of Full-Infrastructure Scenario with Double the Quantity from the Mountain Aquifer, 2010 (lower values) and 2020 (upper values)

in population does not dramatically increase shadow values; however, it sufficiently increases shadow values in Gaza to warrant construction of an additional desalination plant in Rafah in 2020 with an annual capacity of 4 mcm, and the plant in Gaza North is enlarged from less than 1 mcm to 14 mcm per year, as summarized in Table 6.4.10.

Increased agricultural demand

If the amount of water available to agriculture were significantly increased, it is likely that there would be some expansion of irrigated area in Palestine. This is explored by an assumed 30% expansion of the demand curves for agriculture in each of the Palestinian districts. We assume no change in elasticity. The increase in agriculture has a larger impact on shadow values than the parallel increase in population. Gaza is now supplying the West Bank with 12 mcm, leading to shadow values in the West Bank almost all over $1 per cubic meter. Results are similar in 2020 with respect to shadow values, but there is an increased need for desalination (4 mcm in Gaza North and 14 mcm in Rafah), and Gaza supplies the West Bank with 14 mcm, as shown in Table 6.4.10.

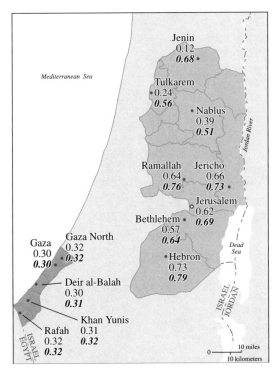

Figure 6.4.9. Comparison of Full-Infrastructure Scenario with Double the Quantity from the Mountain Aquifer and with (upper values) and without (lower values) 150 mcm from Lake Tiberias, 2010

Table 6.4.10. Required Desalination and Pipeline Capacity, Gaza to Hebron, 2010 and 2020 (mcm)

Scenario	Desalination capacity	Quantity transported from Gaza to Hebron
Baseline		
2010	0	<1
2020	<1	0
30% increase in population		
2010	0	0
2020	18	0
30% increase in agriculture		
2010	0	12
2020	18	14
30% reduction of supply due to drought		
2010	22	18
2020	51	27

Drought

Palestine has a highly variable climate, with a fairly high frequency of drought. It is critical that the value of constructing new infrastructure be examined against typical droughts. Droughts of a 30% reduction in annual average supplies are not uncommon, and this case is explored in scenarios for 2010 and 2020. In both years desalination appears to be a good option in Gaza, as shown in Table 6.4.10. In 2010, a total desalination capacity of 22 mcm would be used, increasing to 51 mcm in 2020. A pipeline from Gaza to the West Bank also appears valuable, since it would supply 18 and 27 mcm in 2010 and 2020, respectively.

The drought has a more dramatic impact on water availability and value than either a 30% increase in population or a 30% increase in agricultural demands. Also, these results would be quite different if the supply in the Mountain Aquifer remains at 1995 levels.

7

Results for Jordan

Application of the Water Allocation System model to Jordan, one of the most arid countries of the world, confirms the need for new infrastructure to control leakage and bring additional supplies to urban areas. The model helps evaluate which projects and combinations of projects would be most cost-effective.

1. Background

Jordan is considered a highly water-stressed country, with only 191 cubic meters of freshwater resources available per capita per year. To get a sense of how water stressed Jordan is, it is widely accepted that a per capita freshwater resource of less than 1,700 cubic meters year is considered stressed, less than 1,000 cubic meters is "scarce," and less than 500 cubic meters is "absolute scarcity"—a category comprising only 12 countries (UNEP 2002). Some 90% of the country receives less than 200 millimeters of rainfall per year, and 70% receives less than 100 millimeters per year.

The country may be divided into three physiographic units: the Jordan Rift Valley, the Jordan Highland and Plateau, and the extremely arid south and east Jordan deserts. Significant precipitation, ranging from 200 to 500 millimeters per year, occurs only in the mountains at the eastern side of the Jordan Valley. On average, some 90% of the precipitation is lost to evapotranspiration.

The long-term average availability[1] of fresh surface water resources in Jordan is approximately 500 million cubic meters (mcm) per year. More than half of this quantity occurs in scattered wadis with very irregular discharges. Groundwater is a major water resource in Jordan and the only water resource in many regions of the kingdom. The estimated safe yield of renewable groundwater resources in Jordan is 275 mcm per year. However, many wells are overabstracted; in 2000 the total abstraction from nonfossil groundwater resources was estimated near 525 mcm. An important fossil aquifer (i.e., the last major recharge to this aquifer occurred 10,000 years ago) is located in the Disi area in the south of Jordan. The Water Authority of Jordan estimates the potential capacity of this aquifer at some 125 mcm per year for 50 years. This assumes that Saudi abstraction is limited to current uses and their wells remain at least 50 kilometers from the Jordan border. The water-bearing stratum is a sandstone layer that extends under most of the country's territories, and fresh fossil water has been detected in that layer some 80 kilometers south of Amman. More is yet to be done in exploration of that layer elsewhere before the extent of the country's fresh fossil water is determined.

The main water-using activities in Jordan are agricultural, domestic, and industrial. Although irrigated agriculture contributes to less than 5% of Jordan's national product, it uses about 70% of the water (and somewhat less in drought years). About 50% of the irrigated agriculture is situated in the Jordan Valley. Highland agriculture is either rain fed or supported by groundwater extraction. Because the quality of the water in the Lower Jordan River is poor, the main source of water is the Yarmouk River and the Upper Jordan (Lake Tiberias[2]). Each of these sources supplies the Jordan Valley with water through the King Abdullah Canal. The Lower Jordan has been a significant source since July 5, 1995, through the peace treaty between Jordan and Israel. Sometimes referred to as "peace water," the initial agreed-upon annual quantity is 55 mcm, of which 35 mcm is to be replaced with desalinated water.[3] Additionally, groundwater base flow, mainly upwelling in Wadi Shaq el Barid east of Mukheiba, contributes to the flow of the King Abdullah Canal (18 mcm per year), and further modest quantities of groundwater (20 mcm per year) support irrigated lands not served by the King Abdullah Canal.

More than 85% of the population of Jordan lives in the cities of Amman, Zarqa, Irbid, Mafraq, Jerash, and Ajloun, which are located in the Jordan Highland and Plateau at elevations between 700 and 1,000 meters above sea level—well above the sources of water in the valley. More than 50% of Jordan's population is concentrated in the greater urban area of Amman and Zarqa. With limited local sources of water, domestic and industrial water supply to the capital is a major concern in water resource management, and in summer the supply of sufficient water of acceptable quality to Amman is particularly problematic. About 50% of the total production of water for municipal uses in Amman is unaccounted for—an amount typical for this sector throughout Jordan. This quantity includes leakage, illegal use, unmetered deliveries, and metering and human errors. Several projects are being implemented to improve the situation, but regular short-

Figure 7.2.1. Jordanian Governorates

falls in meeting demands persist, with piped urban water available on fewer than seven days each week in most Jordanian cities.

2. Data for 1995[4]

Districts

Jordan has 12 governorates, which are used as the basis for districts in the model, as illustrated in Figure 7.2.1.

Water consumption data and water demand curves

The procedure used to estimate the demand curves for the three sectors is to estimate a point on the demand curve and assume a price elasticity coefficient. The point on the demand curve is represented by a combination of a price and a quantity of water. For each of the sectors this point was estimated based on data for actual consumption in 1995[5] and the price paid by the consumers of the water in that year.

Table 7.2.1. Urban Water Supply and Consumption Data, 1995

Governorate	Municipal water supply (mcm)	Population	Consumption (mcm)	Consumption per capita per day (liters)	Unpaid-for quantities (mcm)
Amman	89.62	1,696,300	67.22	109	22.41
Zarqa	31.58	687,000	23.69	94	7.90
Mafraq	17.10	191,900	12.83	183	4.28
Irbid	31.97	802,200	23.98	82	7.99
Ajloun	3.48	101,400	2.61	71	0.87
Jerash	3.85	132,500	2.89	60	0.96
Balqa	19.17	301,300	14.38	131	4.79
Madaba	12.89	110,700	9.67	239	3.22
Karak	8.48	182,200	6.36	96	2.12
Ma'an	6.76	85,300	5.07	163	1.69
Tafielah	2.04	67,500	1.53	62	0.51
Aqaba	15.37	85,700	11.53	369	3.84
Total	242.32	4,444,000	181.73	112	60.59

It is critical to distinguish water supply, demand, consumption, and unaccounted-for water. *Supply* is the amount of water that is provided—for example, the amount of water leaving a water treatment plant. *Demand* is the quantity of water desired by a sector at a given price. *Consumption* is the actual amount of water that reaches a given sector. Consumption can differ from demand, primarily when supply is limited or unable to meet demands at the price charged, and therefore the amount consumed is less than the amount demanded at the going price. Water may be *unaccounted for* due to a variety of reasons, but for the purposes of this book, the category is considered water that is lost through physical leakage, or water that is consumed from the system and not paid for (effectively stolen). For example, the unaccounted-for water for the municipal sector is 50–60%, and it is assumed that half of this is due to physical leakage, and half due to unpaid-for water.

The estimated amounts of water supplied, consumed, and demanded for the urban, industrial and agricultural sectors in 1995 are based on data obtained from the Water Authority of Jordan. The demands for 1995 are represented in Tables 7.2.1, 7.2.2, and 7.2.3 for the urban, industrial, and agricultural sectors, respectively. The urban sector is the most complex case because of leakage and unpaid-for water. In all urban districts it is assumed that 25% of supply is lost due to leakage, and 25% is not paid for (Table 7.2.1). In estimating the demand, it is assumed that the demand curve of those not paying for the water is identical to that of those who do pay. This could be problematic if the consumers not paying for water are the very poor, who are not likely to have the same demand curves as more wealthy

Table 7.2.2. Industrial Water Consumption Data, 1995 (mcm)

Governorate	Water consumption
Amman	0.9
Zarqa	6.1
Mafraq	0.3
Irbid	1.0
Ajloun	0.0
Jerash	0.0
Balqa	0.5
Madaba	0.2
Karak	11.7
Ma'an	7.2
Tafielah	5.3
Aqaba	5.8
Total	39.0

Table 7.2.3. Agricultural Irrigated Area and Water Consumption Data, 1995

Governorate	Irrigated area (1,000 ha)	Water consumption (mcm)
Amman	6.8	37
Zarqa	10.6	60
Mafraq	16.9	55
Irbid	11.2	92
Ajloun	1.2	6
Jerash	2.6	12
Balqa	12.4	124
Madaba	0.8	7
Karak	8.2	60
Ma'an	10.6	53
Tafielah	1.8	5
Aqaba	1.9	18
Total	85.0	529

consumers. This does not appear to be the case, however, and we assume that all consumers have the same demand curve on average. (The failure of this assumption would likely affect the elasticity of the demand curve but not its position, and results are not sensitive to reasonable changes in elasticity.[6])

There is a further problem in the data. We have data on consumption, but consumption is not equivalent to demand, which is known to be constrained because of limited supplies (similar to the situation in Palestine, described in Chapter 6). For illustrative purposes only, for 1995 the consumption data are used in place of demand at the prices paid in 1995.

The industrial sector primarily uses groundwater. The amount of pumping is restricted to the amounts shown in Table 7.2.2. As with the household sector, for illustrative purposes, we take these amounts as equal to demand at the price paid by industry of about $1.40 per cubic meter. The fact that such demand must exceed actual consumption only makes our results below stronger.

Agricultural data, including irrigated area and water consumption, are presented in Table 7.2.3. There is no reported unaccounted-for water in this sector. As with the urban and industrial sectors, the quantities consumed are constrained, and therefore the consumption data are not representative of demand. However, for illustrative purposes for 1995, we again begin by using these data.

The average price paid by domestic water users is 38 cents per cubic meter (with adjustments in four of the governorates: 58 cents per cubic meter in the Amman district, 35 cents in Zarqa, $1.18 in Aqaba, and 34 cents in Irbid). Industrial users primarily use groundwater, paying approximately $1.40 per cubic meter. In 1995 agricultural users paid 1.14 cents. To obtain the complete demand curve in the relevant range, we assume constant elasticities: –0.2 for the urban sector, –0.33 for industry, and –0.5 for agriculture, as are generally appropriate for these sectors from the literature.

Supply data

This information is based on data from the Jordanian Ministry of Water and Irrigation for the availability of groundwater and surface water. Jordan has 12 groundwater basins, as illustrated in Figure 7.2.2. Table 7.2.4 gives the annual renewable quantities of water for these basins for each governorate. In addition to the renewable quantities of water, fossil water from the Disi Aquifer was also used in the Ma'an and Aqaba districts in 1995, at a rate of 55 mcm per year and 15 mcm per year, respectively.

Table 7.2.5 gives the data on Jordan's surface water resources. Mean annual discharges are presented for the various rivers and wadis in the 12 districts. Note that flood flows are not now captured. The flow from the Yarmouk River is shown at 246 mcm, as was true in 1995; however, after the construction of dams in Syria, the Yarmouk flows have decreased. This is addressed in the future scenarios. Table 7.2.6 presents the freshwater supply step functions for the districts as they are used in the WAS schematization. Supply step 1 represents the groundwater in a district; supply step 2 represents surface water. The cost estimates of the various supply steps were derived from the Ministry of Water and Irrigation Water Sector Investment Program (1997–2011).

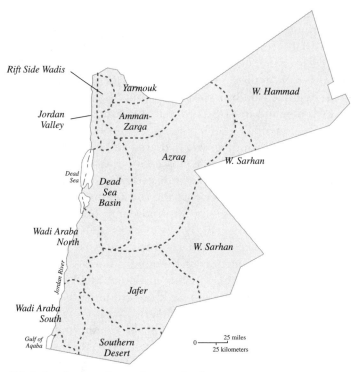

Figure 7.2.2. Jordanian Groundwater Basins

In addition to the production cost of the water, environmental charges have been assumed for the treatment of the effluents produced by households and industry. For urban and industrial water supply, an environmental charge of 30 cents per cubic meter is assumed, as in previous chapters. Plants for recycling water for agriculture existed in all districts in 1995, with capacities as shown in Table 7.2.7.

As in previous chapters, the cost of recycled water for agriculture[7] has been assumed to be 10 cents per cubic meter above the environmental charge, except that in Amman City we assume a lower cost of 5 cents per cubic meter because of economies of scale. Furthermore, we assume that no more then 66% can be recovered from the urban and industrial use of freshwater for use in recycling (as an average feasible figure).

The existing conveyance system

Jordan has an intergovernorate conveyance system that focuses on bringing freshwater to the densely populated area of Amman. Table 7.2.8 presents the interdistrict conveyance systems[8] for each link with the conveyance capacity and costs (in cents per cubic meter). These are the operating costs and do not include capital costs.

Table 7.2.4. Renewable Groundwater Resources by Governorate (mcm)

Governorate	Yarmouk	Amman-Zarqa	Azraq	Jordan Valley	Rift Side Wadis	Dead Sea	Wadi Araba North	Wadi Araba South	Jafer	Southern Desert	Sarhan	Hammad	Total
Amman		12 (14%)	6[a]		2 (13%)	14 (24%)							34
Zarqa		17 (19%)	24[b]		6 (41%)								47
Mafraq	20 (50%)	48 (55%)									1 (8%)	8 (100%)	77
Irbid	18 (45%)			8 (38%)	3 (19%)								29
Ajloun					2 (13%)								2
Jerash	2 (5%)	6 (7%)											8
Balqa		4 (5%)		13 (61%)	2 (14%)								19
Madaba						9 (15%)							9
Karak						16 (35%)							16
Ma'an						3 (6%)	1 (14%)		6 (100%)	<1 (57%)	4 (92%)		14
Tafielah						11 (20%)	1 (35%)						12
Aqaba							2 (47%)	6 (100%)		<1 (43%)			8
Total	40	87	30	21	15	53	4	6	6	<1	5	8	275

Note: Figures in parentheses show the ratio of the area of the groundwater basin to the total area of the governorate. The potential of renewable groundwater is divided into governorates based on the distribution rate of the groundwater basin areas in the governorates.

[a] The distribution area of the A7/B2 aquifer in the Azraq Basin is limited to the southern part of the Amman Governorate

[b] As the renewable groundwater in the central to northern part of the Azraq Basin is mainly abstracted in Zarqa governorate (Al Azraq), and abstraction amount has exceeded its safe yield (24 mcm/yr), all of the safe yield is given to Zarqa governorate.

Table 7.2.5. Historical Average Annual Surface Water Flow by Governorate (mcm)

Governorate	River or wadi	Base flow	Flood flow	Total flow
Amman	None	0.00	0.00	0.00
Zarqa	Wadi Butum	0.00	0.92	0.92
Mafraq	Local wadis	0.00	30.92	30.92
Irbid	Yarmouk River	246.00[a]	109.00	353.00
	Local wadis	32.38	10.89	43.27
	Jordan River	30.00[b]	2.73	32.73
	Subtotal	308.38	122.62	429.00
Ajloun	Wadi Rajib	4.99	1.12	6.11
Jerash	Wadi Zarqa	43.00	25.30	68.30
Balqa	Local wadis	25.18	7.92	33.10
Madaba	Wadis Mujib and Wala	31.38	33.62	65.00
Karak	Local wadis	61.32	12.12	73.44
Ma'an	Local wadis	1.64	17.07	18.71
Tafielah	Local wadis	5.51	1.80	7.31
Aqaba	Local wadis	0.00	2.10	2.10
Total		481.40	255.51	734.91

[a] While Jordan has ownership of 246 mcm per year, only 126 mcm per year is available due to limitations of infrastructure—namely storage facilities to impound floods and a diversion structure to divert the Yarmouk flow to the King Abdullah Canal.

[b] The 30 mcm pertains to the amount that flows from the upper Jordan as a result of the Peace Treaty. While there is no price charged for this water by Israel, there is an additional cost of 4 US cents/m^3.

Table 7.2.6. Annual Renewable Quantities with Fossil Aquifer Use in Step 3, 1995

	Step 1		Step 2		Step 3	
District	Quantity (mcm)	Cost ($/m^3)	Quantity (mcm)	Cost ($/m^3)	Quantity (mcm)	Cost ($/m^3)
Amman	34	0.07	0	—	—	—
Zarqa	47	0.13	0	—	—	—
Mafraq	77	0.13	0	—	—	—
Irbid	29	0.07	158	0.02	30	0.06
Ajloun	2	0.03	5	0.02	—	—
Jerash	8	0.07	43	0.02	—	—
Balqa	19	0.10	25	0.03	—	—
Madaba	9	0.13	31	0.01	—	—
Karak	16	0.07	61	0.01	—	—
Ma'an	14	0.13	2	0.01	55	0.08
Tafielah	12	0.07	6	0.01	—	—
Aqaba	8	0.10	0	—	15	0.08
Total	275		331			

Table 7.2.7. Annual Capacity of Treatment Plants, 1995 (mcm)

Governorate	Recycling capacity
Amman	26
Zarqa	23
Mafraq	1
Irbid	12
Ajloun	2
Jerash	1
Balqa	5
Madaba	1
Karak	1
Ma'an	1
Tafielah	1
Aqaba	2
Total	76

A substantial part of the wastewater from Amman and Zarqa districts is recycled and used for irrigation in the Jordan Valley. The total capacity of the recycling link between Amman-Zarqa and the Jordan Valley (Balqa governorate) is estimated at 65 mcm per year. Because the water is transported by gravity, operation and maintenance costs are low, at 1 cent per cubic meter.

3. Model Results for 1995

Base case results with and without fixed-price policies

The WAS model was run with the data presented above, initially with no fixed-price policies, and constraining withdrawals from supplies to sustainable quantities. A rate of 25% leakage is assumed. The shadow values[9] for this case are shown in Figure 7.3.1.

Note that the Amman district has a relatively high shadow value of approximately 26 cents per cubic meter, versus 2 cents in the neighboring Jerash district. That the difference is greater than the likely operating cost of conveyance between this district and Amman (see Table 7.2.8) indicates a problem stemming from a shortage of conveyance infrastructure.[10] This issue of infrastructure development in Jordan will be explored extensively in the following sections.

Two important aspects of the system as described above are not consistent with how water is currently managed in Jordan: fixed-price policies and the pumping of aquifers beyond sustainable quantities. We begin with fixed-price policies, where there are two items to note.

Table 7.2.8. Freshwater Links with Capacities (upper number) and Costs (lower number), 1995

From	Amman	Zarqa	Mafraq	Irbid	Ajloun	Jerash	Balqa	Madaba	Karak	Ma'an	Tafielah	Aqaba
Amman							1 mcm 1¢/m³	4 mcm 5¢/m³	1 mcm 5¢/m³			
Zarqa	20 mcm 22¢/m³		1 mcm 5¢/m³			1 mcm 5¢/m³						
Mafraq		20 mcm 5¢/m³		5 mcm 5¢/m³		1 mcm 5¢/m³						
Irbid					1 mcm 6¢/m³	1 mcm 6¢/m³	220 mcm 1¢/m³					
Ajloun												
Jerash							1 mcm 1¢/m³					
Balqa	45 mcm 22¢/m³											
Madaba	15 mcm 22¢/m³											
Karak												
Ma'an											1 mcm 2¢/m³	
Tafielah												
Aqaba												

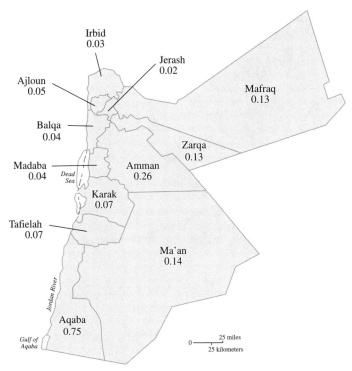

Figure 7.3.1. Baseline Results without Fixed-Price Policies, 1995

First, the existing policies for the urban and agricultural sectors are as presented in the previous section. The second item is more subtle and involves the issue of unaccounted-for water. An estimated 25% of water supplied to the urban sector in 1995 is assumed to have been stolen. Assuming that the characteristics of the users of this water are identical to those of paying users, these users are effectively facing a fixed-price policy of free water for the quantities used. This may be an appropriate policy if these users are extremely poor. It can be handled in WAS by providing this amount of water for free to these users[11] through a set-aside. As depicted in Figure 7.3.2, the striped area represents the value of the water that is provided for free or stolen, the costs of which are borne by society. That value is simply the quantity of unpaid-for water ($Q_{\text{Unpaid-for}}$ in the figure) multiplied by the shadow value of the water (P^*). Unlike the case with fixed-price policies, there is no net loss of social welfare, but rather a transfer of payments within society. In effect, the people taking the water get it for free, but it means the rest of society—effectively the water system—is paying for their water. For 1995, the amount of transfer appears to be on the order of $30 million.

After the free quantities are provided in the model, the price policies described in the previous section are imposed. When this is done for 1995, however, the results show that the policies involved are not feasible without

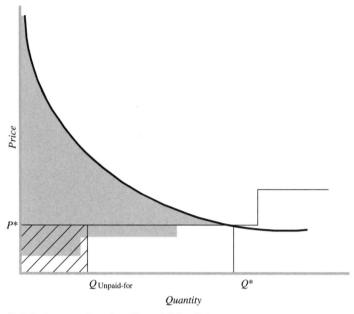

Figure 7.3.2. Accounting for Unpaid-for Water

overpumping of aquifers, which clearly is occurring, as shown in Table 7.3.1. This problem will certainly get worse in the future as demands increase. The infeasibilities are caused by the combination of fixed-price policies and the restriction on using water sustainably—a restriction imposed on all future scenarios. Note that they are not caused by the small quantities of water provided for free. Rather, it is because there is not sufficient water available to Jordan to provide the amount of water demanded at very low fixed prices. This is true now, as evidenced by the fact that water is not available seven days a week, 24 hours a day, even in Amman. Even with the fixed-price policies limited to the quantities of water consumed in 1995, the provision of that quantity of water is not possible without exceeding the annual renewable quantities available. Therefore this type of fixed-price policy is not further explored in this chapter.[12]

4. Projections for 2010 and 2020

Water demand curves

Demand projections have been jointly developed by the Jordanian Ministry of Water and Irrigation and the World Bank for the years 2010 and 2020. One of the scenarios is based on unrestricted demands (rather than on expected water availability), and that is the scenario used in this chap-

Table 7.3.1. Annual Amount of Overabstraction by Groundwater Basin (mcm)

Groundwater basin	Annual renewable quantity of water	Abstraction in 1998	Balance
Yarmouk	40	39	1
Amman-Zarqa	87	150	−63
Azraq	30	52	−22
Jordan Valley	21	36	−15
Rift Side Wadis	15	31	−16
Dead Sea	50	82	−32
Wadi Araba North	4	4	0
Wadi Araba South	6	1	5
Jafer	7	20	−13
Southern Desert	<1	<1	0
Sarhan	5	<1	5
Hammad	8	1	7

ter. Population and demand projections for domestic use are presented in Table 7.4.1. This demand is assumed to be at the same real price paid in 1995 in each governorate.

One further issue with this sector is unpaid-for water. The quantity of water unpaid for in 1995 was presented in Table 7.2.1. This quantity is assumed to grow as the population grows, yielding the quantities presented in Table 7.4.2 by governorate. These quantities will be treated as discussed in the previous section, as a form of fixed-price policy that provides these quantities of water for free.

Although tourism was a relatively small sector in 1995, it is expected to grow in the coming decades, especially along the east coast of the Dead Sea. Estimates of increased water demand for the tourism sector are presented in Table 7.4.3. Despite significant growth, tourism still represents a relatively small portion of the urban demand—on the order of 5%. Most of the water use is characteristic of other urban demands, including basic domestic uses and landscaping. We therefore combine the tourism sector with the urban sector demands in WAS. It is possible, however, that the elasticity of demand for hotels is different from that of households, and probably less elastic. Elasticities can be adjusted through the WAS interface to explore the sensitivity to changes. Just as an example, in this case it could be assumed that hotels are completely inelastic versus households, which have an assumed elasticity of −0.2. Using a weighted average of the two, 95% times −0.2 and 5% times zero would give an adjusted elasticity of −0.19—a difference not worth exploring further.

Industrial water demand projections are based on the existing trend in the growth rate of industry, and an assumption that growth in water use

Table 7.4.1. Population and Domestic Demand Projections, 2010 and 2020

Governorate	2010 population	2010 demand (mcm)	2020 population	2020 demand (mcm)
Amman	2,670,748	116.28	3,517,571	191.21
Zarqa	1,058,505	45.21	1,394,129	75.73
Mafraq	285,471	18.60	375,986	20.60
Irbid	1,257,320	48.99	1,655,982	90.02
Ajloun	171,870	6.55	226,365	12.30
Jerash	207,359	7.79	273,107	14.85
Balqa	466,429	22.81	614,321	33.40
Madaba	168,077	11.14	221,370	12.10
Karak	284,266	12.55	374,400	20.37
Ma'an	135,826	7.45	178,892	9.74
Tafielah	112,336	4.16	147,955	8.04
Aqaba	151,792	10.97	199,922	10.96
Total	6,970,000	312.49	9,180,000	499.30
Annual growth rate	3.1%		2.7%	

Table 7.4.2. Unpaid-for Water Projections, 2010 and 2020 (mcm)

Governorate	2010 Unpaid for	2020 Unpaid for
Amman	35.28	46.46
Zarqa	12.16	16.02
Mafraq	6.36	8.38
Irbid	12.53	16.50
Ajloun	1.47	1.94
Jerash	1.51	1.98
Balqa	7.42	9.77
Madaba	4.89	6.44
Karak	3.31	4.36
Ma'an	2.69	3.54
Tafielah	0.85	1.12
Aqaba	6.81	8.96
Total	95.27	125.48

will grow in parallel. These projections are shown by governorate in Table 7.4.4, with the assumption that these are the demands at the same real prices paid for water in 1995.

The Ministry of Water and Irrigation has developed projections for irrigation water demands for 2010 and 2020 based on several factors, including irrigation patterns in 1995, cropping patterns, distribution of irrigation

Table 7.4.3. Tourism Demand Projections, 2010 and 2020 (mcm)

Governorate	2010	2020
Amman	1.79	3.98
Zarqa	0.02	0.06
Mafraq	0.01	0.01
Irbid	0.04	0.09
Ajloun	0.00	0.01
Jerash	0.00	0.01
Balqa	7.29	7.29
Madaba	6.56	6.60
Karak	0.02	0.05
Ma'an	0.23	0.51
Tafielah	0.00	0.01
Aqaba	1.21	2.69
Total	17.18	21.28

Table 7.4.4. Industrial Demand Projections, 2010 and 2020 (mcm)

Governorate	2010	2020
Amman	1.54	2.41
Zarqa	24.70	30.67
Mafraq	0.40	0.63
Irbid	9.22	10.43
Ajloun	0.00	0.00
Jerash	0.00	0.00
Balqa	0.80	1.26
Madaba	0.31	0.48
Karak	33.14	70.99
Ma'an	12.56	19.69
Tafielah	9.10	14.27
Aqaba	9.81	17.83
Total	101.57	168.66

methods, types of distribution systems, salinity of irrigation water, and climatic conditions.

The resulting total irrigated areas and water demands for each governorate are presented in Table 7.4.5. As in the case of urban and industrial demands, these are the quantities assumed to be demanded at the same real prices paid per unit of water in 1995.

Table 7.4.5. Annual Agricultural Demand Projections, 2010 and 2020

| | 2010 | | 2020 | |
| | Irrigated area | Demand | Irrigated area | Demand |
Governate	(1,000 ha)	(mcm)	(1,000 ha)	(mcm)
Amman	6.8	55	6.8	54
Zarqa	10.6	97	10.6	93
Mafraq	16.9	122	16.9	120
Irbid	17.9	170	17.9	158
Ajloun	1.2	9	1.2	8
Jerash	2.6	23	2.6	20
Balqa	25.8	336	25.8	329
Madaba	0.8	4	0.8	4
Karak	9.4	63	9.4	60
Ma'an	10.6	85	10.6	83
Tafielah	1.8	18	1.8	16
Aqaba	2.3	20	2.3	20
Total	106.5	1002	106.5	963

Baseline runs for 2010 and 2020

The data presented in the previous section were entered into WAS for 2010 and 2020. An initial scenario for both years, without fixed-price policies in place and no changes in infrastructure after 1995, reveals an increasing problem in several governorates, particularly in Amman and Zarqa, as shown by the high shadow values in Figures 7.4.1, and especially Figure 7.4.2. By 2020, this reaches crisis proportions in Amman, Zarqa, and Ajloun, with the shadow value in Amman exceeding $30 per cubic meter.[13] Such shadow values of water are clearly unacceptable and, with neighboring districts at much lower shadow values, indicate a strong need for infrastructure improvements.

One such improvement is suggested by the fact that, by 2020, the shadow value in Aqaba—more than $8 per cubic meter—greatly exceeds even current estimates of the cost of seawater desalination, about 60 cents per cubic meter.

But the main infrastructure problem plainly involves getting water to the capital. There was, in 1995, a conveyance pipeline taking 45 mcm of water from Balqa to Amman per year. Were no further infrastructure to be built, the shadow value in Amman would exceed $30 per cubic meter by 2020, yet the shadow value in Balqa, in the Jordan Valley, would be only about 16 cents per cubic meter, and shadow values elsewhere in the Jordan Valley would be lower still. Plainly, the capacity of that pipeline will not be sufficient by 2020. Hence, either that pipeline must be expanded or other ways found to supply the capital.

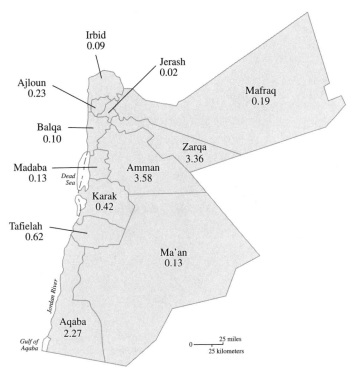

Figure 7.4.1. Baseline Results without Fixed-Price Policies, 2010

Note that this is not a problem of water ownership but a problem of infrastructure. The shadow value of water ownership remains relatively low in the Jordan Valley despite the enormous shadow value in Amman. Since it is always the case in the optimum solution of the model that, for conveyance from A to B,

$$p_B = p_A + t_{AB} + \lambda_{Con} \tag{7.4.1}$$

where

> p_B = the shadow value at B,
> p_A = the shadow value at A,
> t_{AB} = the operating cost of conveyance per cubic meter between the two points, and
> λ_{Con} = the shadow value of conveyance capacity.

The shadow value of the capacity constraint on the Balqa-Amman pipeline must be $33.36 per cubic meter of annual capacity in the run shown in Figure 7.4.2.[14] This is the rate at which systemwide benefits would increase per cubic meter of additional conveyance capacity.

There is an illuminating story related to this. In 1994, when the project was in its infancy, one of the present authors, Munther Haddadin, being

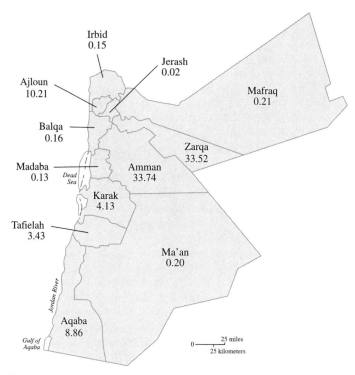

Figure 7.4.2. Baseline Results without Fixed-Price Policies, 2020

exposed to the proposed methods for the first time, asked the rhetorical question, "If the two of us were lost in the desert east of Amman, what then would be the value of a bottle of water?" The answer is that the value of water in the desert would be very high indeed, but the value of water in the Jordan River would not change as a result. In such a case, what is involved is a shortage of infrastructure to convey the water from the river to the desert, not a shortage of ownership of the resources.

Averting the coming crisis: Infrastructure projects and their timing

Not surprisingly, therefore, the Jordanian government has plans to expand the Balqa-Amman pipeline from 45 mcm to 90 mcm per year no later than 2005. In addition, the Zarqa Ma'in project would bring 35 mcm per year of desalinated brackish water from the Balqa district at a cost of 47 cents per cubic meter; it is planned to start in 2006.[15] Beginning with these changes in infrastructure, we see the immediate impact on Amman in terms of shadow values, which drop from $3.58 to 47 cents—the cost of the Zarqa Ma'in project water—in 2010, and from $33.74 to $3.34 in 2020 (Figure 7.4.3). Note that shadow values increase along the Jordan Valley from which the

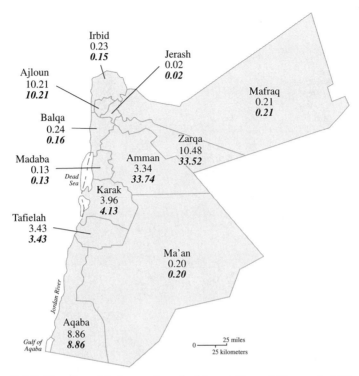

Figure 7.4.3. Results with (upper) and without (lower) Expanded Pipeline from Balqa to Amman Increased from 45 mcm/year to 90 mcm/year and Zarqa Ma'in Project Completed, 2020

water is withdrawn. Irbid sees an increase from 15 to 23 cents, and Balqa from 16 to 24 cents. The scarcity of water in these governorates increases with the increased competition for water. With an overall gain in social welfare of approximately $67 million per year in 2010 and reaching more than $1 billion per year in 2020, the pipeline and desalination projects are clearly essential infrastructure.

In all later runs in this chapter, we assume the capacity of the Balqa–Amman pipeline to have been expanded to 90 mcm per year, and the Zarqa Ma'in project to have been constructed.

In addition to the increased pipe capacity from the Jordan Valley (Balqa) to Amman, another approach to alleviate the crisis would be to reduce intradistrict leakage. The government of Jordan already has plans to bring leakage down to levels of 15% by 2010, with no further improvement expected by 2020. Although this reduction in leakage clearly lowers shadow values in the crisis governorates, the values are still quite high in half the governorates, particularly for Zarqa, Aqaba, and Ajloun, in 2020, as shown in Figure 7.4.4. The gain in social welfare from this reduction in leakage in 2010 is on the order of $41 million per year, and with a discount

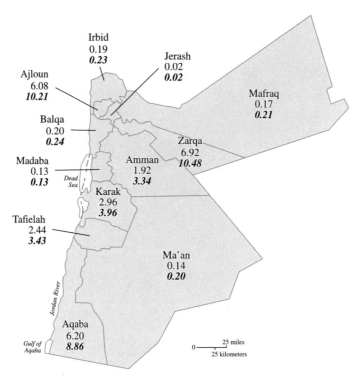

Figure 7.4.4. Results with (upper) and without (lower) Intradistrict Leakage Reduced to 15%, 2020

rate of 5% and a 20-year lifetime, assuming these benefits to be constant (although they clearly will increase over time with increasing demands), the net present value is approximately $520 million. However, the *annual* net benefit is almost $220 million in 2020, suggesting that this could be a critical investment for Jordan over time.

The shadow values in several governorates adjacent to those in crisis are much lower—for example, Irbid, Balqa, Jerash, and Madaba—relative to Amman and Ajloun. This suggests the possibility of interdistrict conveyance. There are social limitations to these transfers, however, in that agriculture in these governorates is of great social importance—for employment as well as for the aesthetic and cultural values associated with agriculture.

The Jordanian government is instead planning to use water from the Disi fossil aquifer to address the problem of persistent water shortages. As described earlier, pumping from this aquifer at a rate of 125 mcm per year is possible for a period of 50 years. A total of 70 mcm is currently being used from this system. An additional 55 mcm per year is added to the system in a new scenario, as well as a pipeline to Amman (initially of unlimited capac-

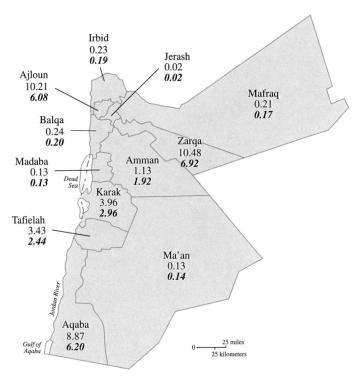

Figure 7.4.5. Comparison of Results (1) with Leakage Reduced to 15% and without Disi Pipeline to Amman (lower) and (2) without Leakage Reduction but with Disi pipeline to Amman (upper), 2020

ity, to let WAS determine the optimal size). Conveyance costs are estimated at $1 per cubic meter.[16]

The results for 2020 are shown in Figure 7.4.5 and compared with the results when leakage is reduced to 15%. The Disi pipeline does not get used at all in 2010 but very clearly alleviates the crisis in Amman in 2020 more than does reducing leakage alone. However, the shadow values in Amman remain quite high, as do the shadow values in Aqaba, Zarqa, Tafielah, Karak, and Ajloun. The net *annual* benefits for the Disi pipeline alone are approximately $40 million in 2020. With a discount rate of 5% and a 20-year project life, the net present value is about $500 million (which should be compared with the capital costs of the pipeline, estimated to be about $600 million). Of course, this assumes no increase in population after 2020, so the actual net benefits are presumably higher.[17] Reducing leakage to 15% will have immediate impact on Jordan's social welfare, largely because the reduction in leakage permits a net gain of 10% more water (since baseline leakage is 25%) throughout the country, whereas the Disi pipeline addresses one district's needs only. The volume of water that

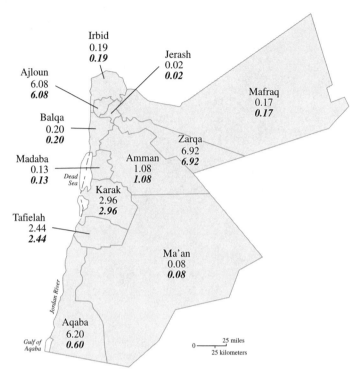

Figure 7.4.6. Comparison of Results with Leakage Reduction and Disi Pipeline to Amman and with (lower) and without (upper) Desalination in Aqaba, 2020

would efficiently flow through the new pipeline, according to our results, is almost 40 mcm per year in 2020.

The combination of reduced leakage and the Disi pipeline, compared with reduced leakage alone, gives an increase in social welfare in 2020 of approximately $10 million per year and further alleviates some of the high shadow values, as shown in Figure 7.4.6 (relative to Figure 7.4.5) for 2020. For example, the per cubic meter shadow values in Amman drop from $1.13 to $1.08, in Zarqa from $10.48 to $6.92, and in Aqaba from $8.87 to $6.21. These values are still high, however. The shadow value in Aqaba in particular, at $6.21 per cubic meter, remains much higher than the existing cost of desalination, suggesting desalination as a solution for that district. To explore this, a desalination plant is added at a cost of 60 cents per cubic meter, with unlimited capacity, thus reducing the shadow value for Aqaba as illustrated in Figure 7.4.6. The net annual benefits are some $30 million, and the required plant capacity is 18 mcm.[18]

There is a longstanding proposal to construct a canal with an annual capacity of 850 mcm from the Red Sea to the Dead Sea (the "Red–Dead Canal"). The difference in elevation would be used to generate electricity, which could in turn be used to desalinate the salt water from the Red Sea, and (after pumping) provide much-needed freshwater to northern Jordan. In addition to providing additional freshwater, the project would make it possible to stabilize the level of the Dead Sea. This would be environmentally beneficial and could enhance tourism and therefore the economy in the region.

It is worth examining whether this development makes sense from the perspective of water needs in Jordan, with a particular focus on Amman, the district with the highest concentration of population. In 2020, with the increased supply of water from the Jordan Valley to Amman, a new pipeline from Disi to Amman, and leakage reduced to 15%, the shadow value in Amman remains at a relatively high $1.08 per cubic meter.

Assuming the Red–Dead project would deliver freshwater to Madaba, the water could then be transferred to Amman at 22 cents per cubic meter.[19] As long as the marginal costs of the desalination did not exceed the difference of $1.08 and $0.22, or $0.86, the project would likely be beneficial from the standpoint of social welfare. The project would likely lower the costs of desalination because the required energy—a major component—would come from the canal's own hydropower plant. But using current costs (estimated at 60 cents per cubic meter, inclusive of capital costs), shadow values in Amman could drop below 80 cents per cubic meter.

Note, however, that this assumes that the Red–Dead Canal is to be undertaken for reasons other than solely the production of desalinated water. The capital costs of the canal itself should therefore not be attributed (or wholly attributed) to the desalination part of the project.

Furthermore, the Red–Dead desalination project would make it inefficient to transfer water from the Disi Aquifer to Amman, since the difference between the two shadow values would not justify the transfer costs ($1 per cubic meter). This does not mean, however, that building the line from the Disi Aquifer would not be valuable if the Red–Dead Canal is to be constructed—quite the contrary. The transfer from Disi may well be needed between 2010 and 2020 while the more complex and time-consuming Red–Dead project can be approved and constructed.

Finally, our infrastructure analysis does not take into account any continuation of Jordan's policy of subsidizing water for agriculture in the Jordan Valley. This really makes no difference so far as averting the coming crisis in Amman is concerned, and we shall examine the effects of such policies in the next chapter.

5. Conclusions

The analyses in this chapter indicate a clear need for infrastructure in Jordan. This is hardly news, and there are extensive infrastructure plans in place in Jordan at this time. However, the results also show that the types of infrastructure that could be constructed are very diverse, and more critically, interrelated. The timing of new infrastructure plays a key role in the value of other infrastructure. This is particularly true in the case of the Disi pipeline to Amman and the proposed Red Sea–Dead Sea project. If the Red–Dead Canal were to come online immediately, the Disi pipeline to Amman would not be necessary, but since the canal will take a very long time to construct, even after all parties have agreed, the Disi pipeline will in the interim provide much-needed water for the city and governorate of Amman.

WAS allows easy examination of these interdependencies of timing and types of infrastructure change, combined with social policies and their impacts. Further, as in the case of desalination facilities in Israel (see Chapter 5), WAS can be used to investigate the optimal capacity to which a given piece of infrastructure should be built.

But beware: it is tempting to suppose that one can investigate this question for all infrastructure projects at the same time by releasing all capacity constraints and recording the throughput for each project in the optimal solution. This is not a good idea, for two related reasons. First, the results of this chapter show very clearly that the desirability of a particular infrastructure project depends on what other infrastructure projects are in existence or being constructed. Second, whether to build a particular project depends on whether the present value of the benefits exceeds its capital costs. In the case of Israeli desalination plants, this did not matter, since capital costs per cubic meter are included in the estimated cost of desalination—60 cents per cubic meter. In a more general investigation, the procedure of relaxing all capacity constraints effectively assumes that all projects or expansion thereof will meet the benefit–cost test and be constructed.

A multiyear version of WAS would solve this problem and permit the optimal timing, order, and capacities of infrastructure projects to emerge as results of the optimization process.

Appendix: Interseasonal Agricultural Water Allocation System

This appendix[20] introduces a linear programming optimization model for analyzing interseasonal allocation of irrigation water in quantities and qualities and their impact on agricultural production and income. The Interseasonal Agricultural Water Allocation System, or SAWAS, is based on the Agricultural Submodel (AGSM), described in Chapter 3. SAWAS can also be considered an application of the main methods of the Water Allocation System (WAS) within the agricultural sector. Like WAS, SAWAS takes demand considerations as well as cost factors and availability constraints into account.

In this research, we stress water scarcity as a problem that arises when water is not found in proper quantity and quality at the appropriate place and time. The model is designed to serve as a decisionmaking tool for planners of agricultural production on both the district and the regional level. It generates an optimal mix of water-demanding activities that maximizes the net agricultural income of the districts and gives the water demands under various prices. It also provides the planner with tools to carry out what-if experiments and generate optimal water demand curves. A principal feature of SAWAS is the use of demand and the benefits from water together with costs and optimization within the agricultural sector to specify the optimal usage of different water qualities. Hence the agricultural planner can use the outputs of SAWAS to bridge the gap between limited water resources and increased agricultural production in an area that suffers from severe water scarcity. The appendix applies the SAWAS model to the Jordan Valley in Jordan.

A.1. Introduction

Water is an essential factor in agriculture and plays a decisive role in economic growth and development. Water resources are generally allocated on the basis of social criteria for maintaining community welfare by ensuring that water is available for human consumption, sanitation, and food production. Societies have invested capital in infrastructure to maintain this allocation. The demand for water and ability to control its location, timing, quality, and quantity are becoming critical with the growing demand for municipal, industrial, and agricultural uses in the arid region.

The basic principle of treating water as an economic good is to allocate it to its best use. After outlining the economic principles behind allocating scarce water resources, we review the actual means of various mechanisms used for allocating water, including marginal cost pricing, social planning, user-based allocation, and water markets. Examples from experience in several countries show that no single approach is suitable for all situations.

Clearly, governments play an important regulatory role, but how effectively they do so depends on the relative political influence of various stakeholders and segments of society. User-based allocation is generally more flexible than government allocation, but collective action is not equally effective everywhere; it is most likely to emerge where there is strong demand for water and a history of cooperation. The outcome of market allocation depends on the economic value of water for various uses, but moving toward tradable property rights in water may ease the process of intersectoral reallocation by compensating the "losers" and creating incentives for efficient water use in all sectors.

A.2. Background

The Jordan River flows to the west of Jordan. Over the centuries the Jordan River has carved a fertile valley that is used for significant production of Jordan's agricultural fruits and vegetables. The river itself is formed from its tributaries—including the Yarmouk River, which flows from Syria, and the Zarqa River, whose headwaters arise entirely from the east bank of Jordan.

At the northern end of the valley, the Jordan River is approximately 200 meters below sea level and drops to more than 400 meters below sea level where it empties into the Dead Sea at its southern end. The valley is approximately 105 kilometers long and 4 to 16 kilometers wide (Jordan Valley Authority 1997, 21). The Jordan Valley is hot during summer and warm during winter. Mean maximum temperature is 36°C during summer, while mean minimum temperature during winter is 14°C, and mean annual 24°C. The mean relative humidity during winter is 65% and 45% during summer. The rainy season is November to mid-April, and rainfall decreases in amount from the northern part (377 millimeters) toward the southern part (87 millimeters). The total water supply of the Jordan Valley averages 252 mcm/yr, the major sources being the Yarmouk and Zarqa Rivers, which historically have supplied the valley with about 72% and 12% of its water, respectively. In addition, the wadis on the east bank of the Jordan River contribute about 16%. The Jordan Valley Authority has developed irrigation in the Jordan Valley and now covers 31,200 ha.

The existing storage and diversion development projects in the valley are the Mukheiba wells, which provide the system with 24 mcm; the Wadi Araba Dam in the northern part of the valley, with a gross storage capacity of 20 mcm; and the Ziglab Dam, with a gross capacity of 4.3 mcm. In the middle part of the valley are two main dams: the King Talal and Karamah Dams, with gross storage capacities of 86 and 53 mcm, respectively. In the southern part of the valley are the Shueib and Kafrin dams, with a gross storage capacity of 2 and 8 mcm, respectively. Finally, the main conveyor of water in the valley is the King Abdullah Canal. This concrete-lined canal extends 110 kilometers along the length of the valley and serves most of the valley's irrigable lands. The intake of the canal is at the mouth of the Yarmouk River, at the northern end of the valley.

A.3. The Problem

Traditionally, the Jordanian government has played a dominant role in managing water resources, but there has been inefficient use of water and poor cost recovery for operating and maintenance expenses (Jordanian Ministry of Water and Irrigation 1993). The mounting cost of developing new water sources and problems with the quality of service in agency-managed systems have led to a search for alternatives to make water allocation and management more efficient.

A water problem arises when water is not found in proper quantity and quality at the appropriate place and time (Kneese and Sweeney 1985, 466). Water scarcity is the single most important natural constraint on Jordan's economic growth. Rapid increases in population and industrial development have placed unprecedented demands on the resource. Actual water uses have exceeded the available Jordanian supply. Renewable groundwater in the north is being exploited at unsustainable rates, and water quality is deteriorating. Existing demands are not being satisfactorily met, and development costs for the remaining resources are rising rapidly.

The total quantity of water consumed in Jordan for all purposes in 1999 was about 950 mcm. Assuming a reasonable growth rate in population, competition between the urban and agricultural sectors for water will intensify, making it essential to optimize water allocation in the agricultural sector. However, water consumption by agriculture depends on several factors, including the technology used, availability of water, and cropping patterns, as well as the amount of land suitable for cultivation.

A.4. Methodology

This appendix demonstrates the interseasonal agricultural water allocation system (SAWAS) as a regional agricultural water allocation system that can be used to allocate scarce water resources among water-consuming activities. SAWAS can serve as a decision support device suggesting to planners what crop patterns are likely to prove optimal under various conditions and relating these to different water policies.

We concentrate on deriving the demand for water and related matters rather than on the optimal cropping pattern, using the Jordan Valley region as a case study. Our model can be used as a general model for allocating agricultural water among districts and regions as well as among seasons where water storage facilities are available.

SAWAS has three main goals: (1) to provide regional and district-level planners with a decision support tool for planning agricultural production under various water amounts, qualities, timing, and prices; (2) to provide the national planner with a tool aimed at regulating the supply side according to certain activities by using storage-transfer; and (3) to provide decisionmakers with a tool suitable for what-if questions of water pricing policy.

SAWAS is a linear optimizing model of agriculture. It uses data on available land, water requirements per unit of land area for different crops, and net revenues per unit of land area generated by the growing of those crops in different locations. These net revenues do not include payments for water, which are handled separately. The model takes prices and quantity allocations for water and generates the cropping pattern that maximizes net agricultural income at the regional level. By varying the price of water of a specific quality while keeping prices of other water qualities constant (*ceteris paribus*), one can construct the demand function for that quality and its impact on water demand for other water qualities and general water demand.

The model can also be used to examine the effects of water quantity allocations among crops and among locations as a result of changes in the prices of agricultural outputs or water restrictions on specific crops and locations.

Objective function. SAWAS is formulated at both the regional level and the district level. Its objective is to maximize net income by selecting the optimal mix of water-consuming activities (field crops, vegetables, and fruit trees). The decision variables are the land areas of the activities. Each activity is characterized by its water requirements per dunam and the net income it produces per dunam, not including water payments. These activities are allocated according to their location in different districts and according to the various qualities of irrigation water.

The objective function that is maximized in SAWAS (equation 7.A.1) is the total annual net income of agriculture in the district. Net income is considered in two parts. The first is what was referred to as water-related contribution. The WRC of activity j is defined as the gross income generated by activity j per unit area less all direct expenses (machinery, labor, materials, fertilizers, etc.) associated with doing so, except for direct payments for water. It measures the maximal ability of the activity to pay for water.

The second component of net income consists of direct payments for water in each activity; its value is subtracted from its corresponding WRC. Such payments do not include water-related expenses because these expenses are included in the price charged by the government for the different water qualities to cover the operational and maintenance costs of the conveyance and distribution irrigation system.

Constraints. The SAWAS model has two main constraints: water and land constraints. The quantity of irrigation water is allocated according to season, location, and quality for each activity. The resulting cropping pattern should at least satisfy the domestic demand, and no enormous change in the actual cropping pattern is allowed by use of constraints on areas.

WATER CONSTRAINTS. The first step is to determine the net water requirements of agriculture in the Jordan Valley. We took into consideration the fact that the efficiency of the irrigation system was 72%, as estimated by the Jordanian Ministry of Water and Irrigation (1999), and that rainfall is 80% effective—that is, 80% of precipitation during the winter season contributes to water crop requirements. (The evaporation and leakage rates are included in irrigation efficiency.)

The water requirements for a given activity increase as we go from north to south because of an increase in temperatures. Each activity can, in principle, use one or more water quality—surface, brackish, and recycled—for the four seasons: fall (October–November), winter (December–February), spring (March–May), and summer (June–September). In addition, SAWAS allows 4 water prices under each water quality, making 12 water prices. Thus the model user can choose a pricing policy for water, allowing prices to perform any required rationing of supplies.

In addition, there is a conditional constraint. An activity that uses brackish or recycled water must mix that water with surface water in a ratio of three to one to meet the irrigation water quality standards required by the Ministry of Water and Irrigation. We impose this requirement as part of the water requirements of agriculture: an activity that uses a certain amount of recycled or brackish water per dunam is assumed also to require one-fourth that amount of surface water.

The second step is the formulation of the supply and the storage-transfer constraints. This was done to specify that current inflows can be either used for current irrigation or stored for later seasons (equation 7.A.2). By contrast, in the original version of AGSM, surplus water in one season could not be used in later ones. The model can be further developed for water transfer among districts or regions. (In this respect, the use of the Jordan Valley example may be special because of the zero cost of using gravity flow in the conveyance system of the King Abdullah Canal.)

LAND CONSTRAINTS. In SAWAS, land areas can be specified according to district, crop groups, and water qualities. This ensures that the sector will produce sufficient quantities of all current crops to meet domestic demand and export activities, and thus the retail prices of crops will be generally stable for the coming season. In AGSM, land area is specified for one district. In SAWAS, land area is divided into three districts, north, middle, and south, and the model optimizes the allocated land areas within the districts and the entire region.

Mathematical notation of SAWAS. The objective function of SAWAS can be written as

$$\max Z = \sum_j \sum_m \sum_k X_{jmk} \times \left[WRC_{jmk} - \sum_i \sum_m (P_{im} W_{im}) - \sum_i \sum_m G_{im} \lambda W_{(i-1)m}^+ \right]$$

$$(7.A.1)$$

where

Z = the maximum achieved total net agricultural income;

X_{jmk} = the total land area of activity j using water quality m in location k;

WRC_{jmk} = the water-related contribution of activity j using water quality m in location k;

P_{im} = the price of water at season i of quality m;

W_{im} = the supply in season i of irrigation water of quality m;

G_{im} = the per cubic meter cost of storing water of quality m in order to transfer it from season $i-1$ to season i;

$W^+_{(i-1)m}$ = the quantity of water of quality m that is so stored; and

λ = the fraction of that water actually available in season i, where $\lambda < 1$ in general because of evaporation and loss during storage. (In our example, λ is assumed to equal 1 because evaporation has been taken into consideration in the calculation of irrigation efficiency in the Jordan Valley and all storage is above ground.)

The objective function is maximized subject to the following constraints:

$$\sum_j \sum_m \sum_k a_{ijmk} X_{jmk} \sum_i \sum_m \left(-W^0_{im} + W^+_{i-1,m} - W^-_{i+1,m} \right) \leq 0 \tag{7.A.2}$$

$$\sum_j \sum_k \sum_n X_{jmk} \leq A_{kn} \tag{7.A.3}$$

where

a_{ijmk} = the water requirement of activity j for water quality m at season i in location k;

W^0_{im} = the total supply of water of quality m in season i, excluding storage;

$W^+_{i-1,m}$ = the transfer of water of quality m from the season before i;

$W^-_{i+1,m}$ = the transfer of water of quality m to the season after i; and

A_{kn} = the total allocated area in location k for crop category n, where the crop categories are field crops, vegetables, and fruit trees.

A.5. Verification and Validation of the SAWAS Model

For purposes of calibration, existing data were used to check the consistency of the results drawn from SAWAS compared with the actual experience in the agricultural sector. The information on the availability of water quality types, the prices paid for the water, the amount of land allocated to each activity by season, crop productivity, total revenue, and WRCs were gathered from various sources in Jordan, as follows:

• the net water requirements of activities in the northern, middle, and southern districts of the Jordan Valley, taking into consideration irrigation efficiency and rain-fed components (the Ministry of Water and Irrigation);

- water supply quantities as an average of 10 years' supply and prices for each water quality (the Ministry of Water and Irrigation);
- current cropping patterns and the blending ratios between surface and recycled or brackish water (the Ministry of Water and Irrigation);
- the WRC figures, estimated on the basis of field surveys carried out in the study area for each crop in the different districts, and farm-gate prices (the Ministry of Agriculture and the Agricultural Market Organization); and
- areas planted in each district (the Ministry of Water and Irrigation and the Department of Statistics).

Several runs and experiments were done to optimize water allocation within the districts and in the region as a whole according to the actual land areas. To be confident of the results obtained from SAWAS, two broad methodologies were used: verification that the matrix was consistent with the problem definition (an inconsistency might be due to incorrect placement of a coefficient), and validation that the problem definition was correct and appropriate. Checking procedures for model consistency were carried out on the final computer results to ensure that the model output compared correctly with its design criteria. However, for this purpose, the entire region was chosen as an example that will be discussed in detail below. Numbers were checked against the original data sources. The result shows that the optimal cropping pattern obtained by SAWAS exceeds the actual cropping pattern in terms of total cultivated area.

In the calibration runs, we used actual 1999 figures for the right-hand-side values of the constraints for water amounts and total land area. Land areas of citrus and banana crops are predetermined because in the short term, these land areas are fixed. The land areas for annual crops, such as field crops and vegetables, have more flexibility in the short term. Therefore, we permitted deviations from the current situation up to 20% from the 1999 data so that we could examine the estimated cropping pattern in the short run and compare it with the optimum solution. The prices per cubic meter of several types of water in 1999 were 4.9 cents for surface water, 0.9 cents for brackish water, and 1.3 cents for recycled water. However, the price of water could be allowed to vary with the season for each kind of water. As an average for all seasons, water storage cost per cubic meter was 0.02 cents.

The results of the validation runs showed a satisfactory fit with the 1999 actual mix of activities and the use of land and water, as can be seen from Table 7.A.1. In this case the model's optimal total water use was 6.93% less than actual values, accompanied by an increase of total cultivable area of 3.07%. Most of the difference comes from area allocated in the south-

Table 7.A.1. Actual and Calculated Data for Jordan Valley

	Actual 1999	*Model results*	*Percentage difference*
Northern area (ha)	11,812	11,812	0.00
Middle area (ha)	7,451	7,451	0.00
Southern area (ha)	5,595	6,357	13.62
Total Jordan Valley area (ha)	24,859	25,621	3.07
Surface water (mcm)	163	163	0.00
Brackish water (mcm)	31	14	–56.17
Recycled water (mcm)	58	58	0.00
Total water (mcm)	252	234	–6.93

ern district, where the model increased the cultivated area with a smaller amount of water. Since the model is an optimizing tool, this is not a large divergence from the actual situation.[21]

A.6. Comparison of the AGSM and SAWAS Models

AGSM is an optimizing linear model for analyzing agriculture given various water quantities and qualities in different seasons. However, in the original version of AGSM, the supply of a certain water quality at a certain season may not be used totally, with some seasons having an excess of water while others have a shortage; there is no provision for the transfer of water between seasons. SAWAS corrects this deficiency, taking storage costs into account.[22] It is interesting to compare the two models' results for the Jordan Valley, using the same technical coefficients. The storage-transfer possibilities in SAWAS make it possible for total water usage to increase from 111.7 mcm with AGSM to 234.1 mcm (110%). Correspondingly, allocated land area increases by 142%, and—perhaps most important—the total agricultural net income of the Jordan Valley increases by 55% (Table 7.A.2).

Table 7.A.3 shows that AGSM has the problem of water scarcity during the spring season where the peak of production takes place together with water abundance in the winter season. This makes the shadow prices[23] of water during the spring season very high. By contrast, in the case of SAWAS, the transfer of excess water quantities from the low demand winter season satisfies the need for irrigation water during the spring and summer seasons.

A.7. Results and Discussion

Allocation of area. The total available planted area already developed in the Jordan Valley is about 31,161 ha, of which 11,812 is in the northern district, 7,451 ha is in the middle, and 11,898 ha is in the southern district. Results of the model show that the optimum allocation of areas was 10,230

Table 7.A.2. General Comparison between AGSM and SAWAS Models

	AGSM	*SAWAS*
Irrigated area		
Jordan Valley (1,000 ha)	10.59	25.62
Northern district (1,000 ha)	85.6	11.81
Middle district (1,000 ha)	0.66	7.45
Southern district (1,000 ha)	1.36	6.36
Total water use (mcm)	111.7	234.1
Surface (mcm)	107.3	139.2
Surface blended with brackish (mcm)	0.1	4.5
Surface blended with recycled (mcm)	1.0	19.2
Brackish (mcm)	0.3	13.6
Recycled (mcm)	3.0	57.6
Total income ($million)	97.1	151.3
Cost of water ($million)	5.3	8.9
Cost of water storage ($thousand)	—	18.4
Total net income ($million)	91.7	142.4

ha planted in the north under surface water, and 1,582 ha used surface water blended with recycled. Moreover, in the middle district, the areas allocated did not use pure surface water at all; rather, surface water was blended with brackish water (91 ha) and with recycled water (7,360 ha). In the southern district all water qualities were used.

Table 7.A.4 shows that about 46% of the valley area was irrigated using surface water, 9% using surface blended with brackish water, and 45% with blended recycled water. The land areas in the northern and middle districts were totally planted; about 5,541 ha in the southern district was not planted because of the shortage in surface water. Accordingly, the land shadow prices were $588.7 per ha and $176.9 per ha in the northern and middle districts, respectively. However, actual land rents were $560, $450, and $280 per ha in the northern, middle, and southern Jordan Valley, respectively.

Allocation of water. About 139 mcm of surface water is used in the optimal solution. Of this, about 70% is used in the northern district because of the high profitability of crops planted there. By contrast, 89% of recycled water is used in the middle and southern districts. In general, about 50% of the total water supply is allocated to the northern district, 14% to the middle district, and 36% to the southern district (Table 7.A.5).

Total net income. Total net income from the optimum cropping pattern was $142.4 million, of which about 79% is generated from surface irrigation

Table 7.A.3. Comparison between AGSM and SAWAS models in water use, transfer, and shadow prices

Water qualities by season	Water supply (mcm)	AGSM Use (mcm)	AGSM Shadow price ($/m³)	SAWAS Use (mcm)	SAWAS Transfer (mcm)	SAWAS Storage cost ($thousand)	SAWAS Shadow price ($/m³)
Surface water	162.9	108.4		162.0	54.0	10.8	0.232
Fall	29.0	21.0	0.0	29.0	0.0	0.0	0.093
Winter	40.7	3.2	0.0	5.0	35.7	7.1	0.093
Spring	22.4	22.4	3.37	39.8	18.3	3.7	0.093
Summer	70.9	61.8	0.0	89.2	0.0	0.0	0.093
Brackish water	31.1	0.3		13.6	8.4	0.6	0.009
Fall	11.0	0.1	0.0	3.5	5.2	0.0	0.009
Winter	7.3	0.0	0.0	1.9	3.2	0.6	0.009
Spring	0.1	0.1	2.74	3.2	0.0	0.0	0.009
Summer	12.7	0.2	0.0	5.0	0.0	0.0	0.009
Recycled water	57.6	3.0		57.6	34.7	6.9	0.070
Fall	12.7	0.6	0.0	12.7	0.0	0.0	0.014
Winter	27.5	0.2	0.0	2.5	25.0	5.0	0.014
Spring	0.5	0.5	3.29	15.7	9.8	2.0	0.014
Summer	16.9	1.8	0.0	26.7	0.0	0.0	0.014
Total	251.6	111.7		234.1	97.1		

Table 7.A.4. Irrigated Areas According to Different Water Qualities (ha)

Water quality	North	Middle	South	Jordan Valley	Percentage
Surface	10,230	0	1,490	11,720	46
Surface blended with brackish	0	91	2,322	2,413	9
Surface blended with recycled	1,582	7,360	2,545	11,488	45
Total irrigated area	11,812	7,451	6,357	25,621	100
Percentage	46	29	25	100	

Table 7.A.5. Water Use According to Water Quality (mcm)

Water quality	North	Middle	South	Jordan Valley	Percentage
Surface	109.05	0.00	30.12	139.17	59
Surface blended with brackish	0.00	0.05	4.49	4.54	2
Surface blended with recycled	2.15	8.32	8.73	19.20	8
Brackish	0.00	0.15	13.46	13.61	6
Recycled	6.45	24.95	26.20	57.60	25
Total water use	117.64	33.46	83.01	234.11	100
Percentage	50	14	36	100	

water and 20% comes from blended recycled water. The effect of blended brackish water on net income was negligible. Geographically, about 67% of total net income comes from the northern district, with about 9% and 24% coming from the middle and the southern districts, respectively. This can be attributed mainly to the existence of citrus in the north and bananas in the south (Table 7.A.6).

Storage and transfer of water supply. The importance of the storage-transfer rows in the SAWAS model stems from the fact that excess water in the fall or winter season can be transferred to the spring and summer seasons in which there is a high demand for water. Table 7.A.7 shows that, in the case of surface water, 35.7 and 18.3 mcm of water were transferred from the winter and spring seasons, respectively, to cover the greater demand in the spring (39.8 mcm) and summer (89.2 mcm) seasons. Without such transfers, the available water supply quantity is only 22.4 mcm in spring and 70.9 mcm in summer. Similarly, the interseasonal transfer of recycled water was 25 mcm and 9.8 mcm from winter and spring, respectively, to cover the shortage that would otherwise occur in spring and summer. In the case of brackish water, the model transferred only 8.4 mcm from fall and winter seasons, and yearly brackish water is in excess supply by 17.4 mcm.

Table 7.A.6. Total Net Income According to Different Water Qualities ($million)

Water quality	North	Middle	South	Jordan Valley	Percentage
Surface	90.7	0.0	22.3	113.0	79
Surface blended with brackish	0.0	0.1	1.1	1.1	1
Surface blended with recycled	4.7	12.7	10.9	28.3	20
Total net income	95.4	12.7	34.3	142.4	100
Percentage	67	9	24	100	

Table 7.A.7. Available Water Supply, Use, and Storage Transfer Quantities and Shadow Prices

Season	Actual supply (mcm)	Transfer to next season (mcm)	Optimal use (mcm)	Shadow price ($/m³)	True economic price ($/m³)
Surface water					
Fall	28.97	0.00	28.97	0.1833	0.2323
Winter	40.67	35.69	4.98	0.0441	0.0931
Spring	22.36	18.28	39.77	0.0441	0.0931
Summer	70.91	0.00	89.19	0.0441	0.0931
Subtotal	162.91	53.97	162.91		
Brackish water					
Fall	10.99	5.17	3.46	0.0	0.009
Winter	7.31	3.24	1.94	0.0	0.009
Spring	0.05	0.00	3.24	0.0	0.009
Summer	12.70	0.00	4.97	0.0	0.009
Subtotal	31.05	8.41	13.61		
Recycled water					
Fall	12.73	0.00	12.73	0.0575	0.0705
Winter	27.46	24.98	2.49	0.0015	0.0145
Spring	0.50	9.77	15.71	0.0015	0.0145
Summer	16.90	0.00	26.67	0.0015	0.0145
Subtotal	57.60	34.75	57.60		

A.8. Deriving Water Demand Curves

After its verification and validation, the model was ready for systematic runs to evaluate the response of water quantity to a wide range of water prices. In these runs the restrictions on the right-hand-side values of the land-area constraints were relaxed. For example, the total potential cultivable area in the Jordan Valley was used instead of the area actually under cultivation in 1999. To obtain observations on the demand curves, we held constant the prices of two of the water qualities while systematically changing the price of the remaining one. Thus, for example, we allowed the surface water price to range from 5 cents to $1.10 per cubic meter, holding the prices of brackish and recycled water constant at their actual prices of 0.9 cents and 1.3 cents, respectively. We allowed the brackish water price to range from 0.5 cents to 5.5 cents per cubic meter, holding other water prices constant (surface, 4.9 cents; recycled, 1.3 cents); and we allowed recycled water price to range from 1 cent to 33 cents per cubic meter, holding other prices constant (surface, 4.09 cents; brackish, 0.9 cents). In these runs, we used the same water price for all seasons for a given water type.

Deriving demand functions for different water qualities. The results of systematically changing the surface water price are presented in Table 7.A.8. The columns of Table 7.A.8 are explained by their respective headings except for the last column (which will be discussed later). For example, the fourth row of the table reads as follows: for surface water price $P_s = \$0.20$ per cubic meter, the entire irrigated area is 23,513 ha, of which 10,155 ha is irrigated with surface water; the total water demand is 214.0 mcm, of which 131.5 mcm is surface water; the total water expenses are $27.27 million, of which $22.18 million are the expenses of surface water. The net income is $119.4 million, of which $73.2 million comes from surface water. The profitability of 1 hectare is $10,245, whereas the profitability of 1 cubic meter is 70.4 cents.

When the price of surface water rises from 20 cents to 25 cents per cubic meter, a reduction in the irrigated area occurs (from 23,513 to 22,052 ha). This is because some crops, like alfalfa, leave the optimal solution because they are no longer competitive with other crops. The quantity of surface water demanded is reduced by 13 mcm. On the other hand, the use of other water qualities, mainly recycled, increases by 6 mcm, partially compensating for the decline in surface water usage. Columns 1 and 5 in Table 7.A.8 are surface water price and total demand of surface water, respectively. They are graphically presented in Figure 7.A.1, creating the optimal demand curve for surface water in the Jordan Valley. A linear approximation to the curve fits quite well; the corresponding regression coefficients are presented in Table 7.A.11.

The own-price elasticity of surface water demand, which is derived from equation 7.A.4a in Table 7.A.11, is about –0.04 at the actual surface water

Table 7.A.8. Responsiveness to Incremental Increase in Surface Water Price

Price ($/m³)	Planted area Total (ha)	Planted area Surface water (ha)	Water use Total (mcm)	Water use Surface (mcm)	Water expenses Total ($million)	Water expenses Surface ($million)	Total net income Total ($million)	Total net income Surface ($million)	Profitability Land ($/ha)	Profitability Water ($/m³)
0.05	25,621	11,720	234.1	162.9	9.02	7.26	142.3	112.8	11,875	0.811
0.10	25,621	11,795	235.2	157.4	16.67	13.80	134.1	102.1	11,152	0.776
0.15	25,023	12,368	233.0	151.7	23.72	19.71	126.4	91.6	10,075	0.735
0.20	23,513	10,155	214.0	131.5	27.27	22.18	119.4	73.2	10,245	0.704
0.25	22,052	8,632	207.1	118.5	30.65	24.08	113.3	59.6	10,303	0.670
0.30	21,811	8,730	206.0	117.4	36.24	28.56	107.5	54.7	10,060	0.622
0.35	21,811	8,730	206.0	117.4	42.11	33.32	101.6	50.3	10,060	0.572
0.40	20,867	8,582	203.5	114.8	46.96	37.07	95.7	44.8	9,938	0.525
0.45	20,867	8,582	203.5	114.8	52.71	41.70	90.0	40.5	9,938	0.475
0.50	19,984	8,148	197.6	109.0	55.51	43.40	84.4	34.2	9,746	0.431
0.55	16,479	7,048	182.7	94.1	52.77	39.55	79.6	25.9	9,155	0.401
0.60	15,735	6,721	179.5	89.6	54.82	40.49	75.0	21.6	8,941	0.359
0.65	14,729	6,244	173.0	83.2	55.11	39.67	70.5	17.3	8,591	0.323
0.70	14,079	4,931	156.8	65.3	46.73	30.16	66.8	12.2	7,240	0.342
0.75	14,056	4,921	156.7	65.1	49.89	32.21	63.6	10.4	7,227	0.293
0.80	11,284	3,115	129.3	40.7	33.56	14.81	60.5	7.6	3,570	0.681
0.85	11,017	1,790	122.9	34.2	30.10	10.23	58.7	6.4	2,600	1.379
0.90	10,879	823	119.5	30.9	28.81	7.83	57.1	3.9	1,598	2.950
0.95	10,879	823	119.5	30.9	30.35	8.27	55.5	3.8	1,598	2.900
1.00	10,879	823	119.5	30.9	31.89	8.70	54.0	3.7	1,598	2.850
1.05	10,879	823	119.5	30.9	33.44	9.14	52.4	3.7	1,598	2.800
1.10	10,825	0	118.2	29.5	33.53	8.13	50.9	0.0	0	0.000

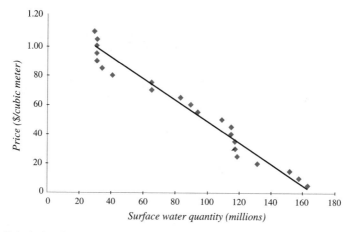

Figure 7.A.1. Surface Water Demand Curve

price of 4.9 cents per cubic meter. This is a very low elasticity, but that is largely a consequence of the very low actual price at which it is evaluated. At the midpoint of the range of surface prices studied (57.5 cents per cubic meter), the own-price elasticity of surface water demand is about –0.91. This means that, starting at that price, an increase of 1% in the price of surface water will decrease the quantity demanded by about 0.91%, so demand is slightly inelastic (Table 7.A.11). Using the same procedure, the total water quantity demanded is regressed on surface water price, holding the prices for brackish and recycled water constant (equation 7.A.4b). The overall water demand elasticity is –0.027 at the actual surface water price 4.9 cents per cubic meter, but –0.42 at the average of 57.5 cents per cubic meter of surface water. This means that increasing the price of surface water by 1% decreases the quantity demanded of all kinds of water by 0.42%.

From equation 7.A.4a in Table 7.A.11, we determined the price level at which the absolute value of the elasticity will be equal to or greater than 1. The water price level at which the price elasticity of water is unitary is 60.45 cents per cubic meter. Above this price, water demand is price elastic and at lower prices, water demand is price inelastic.

Following the same procedure as for surface water, the results of varying the brackish and recycled water prices are presented in Tables 7.A.9 and 7.A.10, respectively. The best-fitting demand curve for brackish water is again linear, as shown in Figure 7.A.2, whereas in the case of recycled water the semilog form fits best (Figure 7.A.3).

The price elasticities of demand of brackish and recycled water, derived from equations 7.A.5a and 7.A.6a in Table 7.A.11, are estimated at –0.29 and –0.43, at the actual water prices of 0.9 cents and 1.3 cents per cubic meter, respectively. The price elasticities of demand of brackish and recycled water, at the respective midpoint prices of 3 cents and 1.7 cents per

Table 7.A.9. Responsiveness to Incremental Increase in Brackish Water Price

Price ($/m³)	Planted area Total (ha)	Planted area Brackish (ha)	Water use Total (mcm)	Water use Brackish (mcm)	Water expenses Total ($million)	Water expenses Brackish ($million)	Total net income Total ($million)	Total net income Brackish ($million)	Profitability Land ($/ha)	Profitability Water ($/m³)
0.0050	25,621	3,085	234.6	31.1	8.73	0.40	142.5	1.6	10,082	0.052
0.0075	25,621	2,736	234.3	24.4	8.81	0.36	142.4	1.3	8,905	0.054
0.0100	25,621	2,413	234.1	18.1	8.87	0.30	142.4	1.1	7,519	0.061
0.0125	25,621	2,413	234.1	18.1	8.90	0.34	142.4	1.1	7,519	0.059
0.0150	25,621	2,413	234.1	18.1	8.94	0.37	142.3	1.0	7,519	0.057
0.0175	25,590	2,399	234.0	18.0	8.97	0.40	142.3	1.0	7,484	0.056
0.0200	25,107	2,145	231.6	14.8	8.95	0.36	142.3	0.9	6,904	0.063
0.0225	25,107	2,145	231.6	14.8	8.98	0.39	142.2	0.9	6,904	0.061
0.0250	25,023	2,109	231.2	14.3	9.00	0.40	142.2	0.9	6,781	0.061
0.0275	25,023	2,109	231.2	14.3	9.03	0.43	142.2	0.8	6,781	0.059
0.0300	25,023	2,109	231.2	14.3	9.05	0.45	142.2	0.8	6,781	0.057
0.0325	25,023	2,109	231.2	14.3	9.08	0.48	142.1	0.8	6,781	0.055
0.0350	25,023	2,109	231.2	14.3	9.11	0.51	142.1	0.8	6,781	0.053
0.0375	24,930	1,963	229.7	12.3	9.08	0.46	142.1	0.7	6,241	0.056
0.0400	24,813	1,801	227.8	9.7	9.02	0.38	142.1	0.6	5,394	0.065
0.0425	24,461	1,203	225.7	6.9	8.95	0.28	142.0	0.4	5,718	0.054
0.0450	24,445	1,170	225.4	6.6	8.95	0.28	142.0	0.4	5,619	0.054
0.0475	24,383	1,040	225.1	6.2	8.95	0.28	142.0	0.3	5,923	0.046
0.0500	24,244	789	224.3	5.0	8.92	0.23	142.0	0.2	6,349	0.035
0.0525	23,921	337	222.1	2.1	8.82	0.10	142.0	0.1	6,349	0.033
0.0550	23,680	0	220.5	0.0	8.73	0.00	142.0	0.0	0	0.000

Table 7.A.10. Responsiveness to Incremental Increase in Recycled Water Price

Price ($/m³)	Planted area		Water use		Water expenses		Total net income		Profitability	
	Total (ha)	Recycled (ha)	Total (mcm)	Recycled (mcm)	Total ($million)	Recycled ($million)	Total ($million)	Recycled ($million)	Land ($/ha)	Water ($/m³)
0.01	25,621	10,077	234.7	61.5	9.60	1.95	141.6	27.0	6,106	0.439
0.03	25,107	9,272	232.4	51.3	10.19	2.39	140.7	23.1	5,531	0.451
0.05	24,923	8,940	230.3	48.5	10.81	2.99	139.9	21.6	5,426	0.446
0.07	22,853	3,143	211.2	23.0	9.82	1.77	139.5	12.0	7,331	0.521
0.09	22,785	2,686	210.0	21.5	10.03	1.97	139.1	10.9	7,987	0.510
0.11	22,133	2,688	202.6	11.5	9.38	1.23	138.9	4.6	4,273	0.399
0.13	21,811	2,205	199.7	9.3	9.29	1.13	138.7	4.1	4,208	0.445
0.15	21,811	1,851	199.7	7.7	9.24	1.05	138.6	3.2	4,165	0.415
0.17	21,811	1,836	199.7	7.6	9.35	1.16	138.5	3.0	4,163	0.399
0.19	21,811	1,836	199.7	7.6	9.47	1.27	138.4	2.9	4,163	0.384
0.21	21,811	1,836	199.7	7.6	9.58	1.39	138.2	2.8	4,163	0.369
0.23	20,867	941	196.9	3.9	8.99	0.76	138.2	1.3	4,108	0.329
0.25	20,867	941	196.9	3.9	9.04	0.82	138.1	1.2	4,108	0.314
0.27	20,867	701	196.1	2.9	8.89	0.65	138.0	0.9	4,108	0.299
0.29	20,867	141	194.4	0.6	8.40	0.14	138.0	0.2	4,108	0.284
0.31	20,776	0	194.0	0.0	8.26	0.00	138.0	0.0	0	0.000

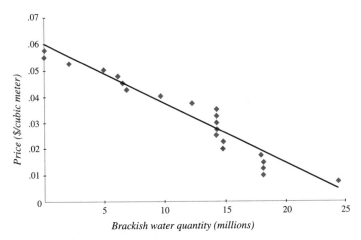

Figure 7.A.2. Brackish water demand curve

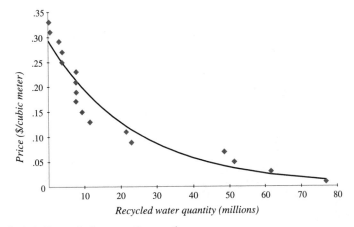

Figure 7.A.3. Recycled water demand curve

cubic meter, are –1.01 and –1.21, respectively, so demand is almost unitary elastic for brackish water and elastic for recycled water (Table 7.A.11).

The price levels at which the absolute value of the price demand elasticity is equal to or greater than 1 were determined. The water price level at which the price elasticity of water demand is unitary is 2.99 cents per cubic meter for brackish water and 11 cents per cubic meter for recycled water.

The effect of increasing brackish and recycled water prices on overall water demand is measured by an elasticity of –0.01 at the actual price for recycled water and –0.06 at the actual price for brackish water. Even at the midpoints of the ranges studied, the elasticity is also small, –0.07 at the recycled water price and –0.03 at the brackish water price.

Table 7.A.11. Demand Functions and Price Elasticities for Different Qualities of Irrigation Water

Equa-tion	Quality of irrigation water	Demand function	R^2	Price elasticity at Actual price	Midpoint
(4a)	Surface	$Q_s = 166{,}465{,}578 - 137{,}680{,}536\,P_s$ $(-22.49)*$	96.2	−0.0414	−0.9068
(4b)	All water	$Q_w = 248{,}185{,}634 - 128{,}289{,}162\,P_s$ $(-20.43)*$	95.4	−0.0269	−0.4229
(5a)	Brackish	$Q_b = 26{,}516{,}202 - 443{,}092{,}452\,P_b$ $(-12.35)*$	89.0	−0.2930	−1.0052
(5b)	All water	$Q_w = 237{,}673{,}983 - 263{,}563{,}132\,P_b$ $(-13.00)*$	89.8	−0.0101	−0.0344
(6a)	Recycled	$Q_r = 30{,}344{,}654 - 24{,}604{,}638\,\ln P_r$ $(-15.44)*$	94.1	−0.4272	−1.2117
(6b)	All water	$Q_w = 177{,}327{,}516 - 14{,}788{,}430\,\ln P_r$ $(-9.3)*$	85.1	−0.0632	−0.0712

Note: Q = water quantity demanded (m³), and P = price of water ($/m³). The subscripts s, b, r, and w stand for water quality: surface, brackish, recycled, and all water, respectively.

* Significant at 1% level

A.9. Conclusion

The application of SAWAS to data for the Jordan Valley suggests that the model closely approximates the actual response of farmers to water prices. This is borne out by the fact that estimates of elasticity of demand for water obtained from SAWAS are reasonable.

Because the results generally agree with actual behavior, they can serve planners as an approximation to the real world. SAWAS also provides a quantitative postoptimal sensitivity analysis that can be used to analyze uncertainty, stability of plans, and risks.

The interseasonal allocation of agricultural water permits the efficient use of water, which in turn plays an important role in increasing total agricultural income in the region. The model can produce insights for agricultural planners, who must allocate scarce water resources among agricultural activities by time, space, and water quality. It also generates estimates of the effects of different water prices. Indeed, we wish to stress that water pricing, aided by analyses such as this, can be an appropriate and efficient means of controlling agricultural water consumption.

Finally, as shown in our case study, SAWAS can be applied to a scale wider than just a single district and also can be developed to be an interregional water-allocating system by accounting for the cost of water transfer between regions.

CHAPTER
8

The Value of Cooperation

The uses of the Water Allocation System are not restricted to domestic management. As discussed in Chapter 4, the WAS tool can also be used in resolving water disputes. We summarize some of that discussion here.

Water and water disputes can be monetized and analyzed in terms of economics, taking full account of water's social or national value, which may exceed its private value. This can assist by showing the true size of the water ownership problem. In the case of Israel, Jordan, and Palestine, the ownership problem is not very large, and this is likely to be true at least wherever seawater desalination is available and conveyance costs from the seacoast modest. Further, each party can use its own version of WAS to evaluate the consequences of different water agreements.

Next, the parties can cooperate by agreeing to trade short-term permits to use water at prices that reflect its scarcity value. WAS generates such prices and shows the gains from cooperation.

Perhaps most important of all, the WAS tool can be used to guide cooperation in water and to estimate what such cooperation would be worth. Basically, WAS-guided cooperation consists of neighbors' trading "water permits"—short-term access to each other's water—and doing so at efficiency prices generated by WAS. Such prices reflect the values put on water by each participating entity. Since trade in water permits is voluntary, both the buyer and the seller of water permits gain from such transactions. The buyer receives water that it values more highly than the money given up to buy it; the seller receives money that it values more highly than the water it gives up in the sale. The result is a win–win situation.

196

1. Cooperation versus Ownership

In this concluding chapter, we examine potential cooperation among the various parties, looking at results drawn from runs of the WAS model. Those runs are of two sorts. First, we consider what regional cooperation would look like in terms of water flows and infrastructure ("regional runs"). Since the question of the optimal use of water has the same answer regardless of the ownership of water, our results here are independent of the answer to the water ownership question.

Second, we investigate the benefits that each party would gain from cooperation and compare these with the benefits stemming from additional water ownership ("gains runs"). Obviously, the answers here depend on assumptions about water ownership, and since we do not presuppose the answer to the ownership question, we have to investigate several cases. We do this most extensively when analyzing bilateral cooperation and varying assumptions about water ownership. Our results must be taken as illustrative only, although they are, in fact, very striking.

There are some subsidiary issues to follow up in the gains runs. In particular, we compare two-party with three-party cooperation and find that when one starts with two-party cooperation between A and B and adds a third party, C, either A or B typically loses rather than gains. On the other hand, that party, say A, is still better off than it would be on its own, and the joint gains to A, B, and C together are always sufficiently large that they can be redivided so that everybody wins. It is easy to show that this result will always hold.

Next, there is the question of shifts in price policy toward agriculture. We investigate whether an agricultural water subsidy by one cooperating party will damage the others, making agricultural water policy a subject for continuing negotiation in the context of a cooperative agreement. We find that, so far as water alone is concerned, this is at most a very minor problem and that, indeed, for some ownership divisions, the nonsubsidizing parties actually gain from a water subsidy put in place by their partner.

Although we do not take a position on the ownership question itself, we do show that it is relatively unimportant, save in symbolic terms. We remarked in Chapter 4 on the use of WAS to examine the consequences of different water quantity agreements, using the case of the Banias and Israel–Syria to illustrate such results. A similar analysis applies to the issue of Lebanese pumping of the Hasbani—an issue that caused considerable excitement a few years ago.

Security issues or other possible objections to the cooperative plan envisaged may yet arise. After reading this chapter, the reader may wish to review the discussion of such issues in Chapter 4.

We recognize, of course, the symbolic and emotional value attached to the ownership of some bodies of water, the Jordan River in particular. We therefore consider the consequences if the owner of such a water source were to withhold its waters from any cooperative arrangement.

Finally, there is (as always) the problem of Jerusalem. We take no position on the resolution of that issue. In Chapter 5, we accepted the Israeli team's view of the world and Israel's responsibility for supplying water to Jerusalem, including East Jerusalem. And in Chapter 6, we accepted the Palestinian team's view of the world and Palestine's responsibility for supplying water to East Jerusalem. This did not matter so long as we were analyzing results for individual countries, but we must now consider what to do with the inconsistency between those views when analyzing cooperative runs.

In the case of the regional runs, the problem of Jerusalem presents no great difficulty. There will be people living in East Jerusalem, and those people must be supplied with water, and it should be done in the regionally efficient way. We handle this mechanically, by connecting the Palestinian team's Jerusalem district and the Israeli team's Jerusalem Mountains district by a conveyance line that costs nothing to operate in either direction. We then arbitrarily subtract the demand of the Palestinian-estimated households in the Palestinian Jerusalem district from household demand in the Israeli Jerusalem Mountains district. This effectively avoids double-counting the population of East Jerusalem and would give the same result as any other assignment of that population. (We avoid presenting schematics with either district on them.)

Two minor difficulties are present. First, the population forecasts for East Jerusalem implicitly made by the Israeli team probably do not match the similar forecasts made by the Palestinian team. Secondly, the elasticity of demand used in the Israeli runs is greater (absolutely) than that used in the Palestinian runs. We ignore the first issue and handle the second by assuming that, under cooperation, future Palestinian residents of East Jerusalem will be sufficiently well off to have the same price-elasticity of water demand as Israelis. In any event, the resolution of these issues hardly affects our results. Indeed, the whole Jerusalem issue can only have a minor effect because the populations and water demands involved are not very large. It would, of course, be easy to investigate such effects in detail, but we do not present such results.

When we come to the gains runs, however, the Jerusalem problem cannot effectively be handled in this way. We want to compare the benefits achieved with and without cooperation, but although the cooperative half of such a comparison could be done by handling Jerusalem as in the regional runs, the noncooperative half cannot. Since we see no way of solving this without at least an implicit assignment of East Jerusalem to either Israel or Palestine, and since we strongly wish to avoid making a political statement, we have decided to live with the inconsistency. Hence in the gains runs—both cooperative and noncooperative—we stay with a double-counted population of East Jerusalem, assigned to both parties. Of course, this affects our quantitative results slightly, making them only illustrative, but the illustration is very powerful.

2. Regional Cooperation

We now examine what regional cooperation would look like in terms of water flows. We do this by optimizing the use of the available water over all three countries together. The optimal solution is independent of the ownership of the water involved. We also examine the case in which Israel and Palestine cooperate without Jordan.

We shall concentrate on the following questions:

- How much Jordan River water would be taken by Palestine in the Jenin District (by means of a pipeline assumed to have been built, with an operating cost of 12 cents per cubic meter)?
- How much water would be supplied to Gaza via the Israeli National Carrier?
- How much treated waste water would be supplied to Negev agriculture from one or more treatment plants in Gaza?
- How much Jordan River water would be taken by Jordan at Irbid?
- What would be the shadow value of freshwater in Amman?
- How much water (if any) would flow to Amman over the conveyance line from the Disi aquifer?
- Would desalinated water from the Red Sea–Dead Sea Canal flow to Amman?
- How much desalination (if any) would be required on the Mediterranean coast?

We examine these questions for both 2010 and 2020, under differing assumptions about infrastructure and policies. The runs of WAS, distinguished by a letter following the year, are described in Table 8.2.1. The results are summarized in Tables 8.2.2 (2010) and 8.2.3 (2020). Each table has two parts—one with Israeli fixed-price policies (FPPs) in effect, and one without such policies.

Throughout, we assume that both Palestine and Jordan will have reduced their intradistrict leakage problems to 20% and 15%, respectively. We also assume that all currently planned infrastructure has been built (with the exception of the Red Sea–Dead Sea Canal, on which we comment separately).[1]

Of course, all our results must be taken as illustrative. Although they are probably qualitatively valid, their quantitative aspects necessarily depend on assumptions and forecasts.

The following observations appear of interest:

Table 8.2.1. Description of Cooperative Runs

Run	Description
A	Full 3-way cooperation. There is a treatment plant in Gaza with conveyance to the Negev. Israeli fixed-price policies are in effect. A conveyance line from the Israeli National Carrier to Gaza can carry freshwater in addition to the 5 mcm that is now so delivered.
B	Same as A, except that there is no conveyance of treated wastewater from Gaza to the Negev.
C	Israeli–Palestinian cooperation only. Total available Jordan River water is 662 mcm (the average annual flow into the lake) less the 55 mcm delivered to Jordan by treaty.
D	Same as A, but with naturally occurring freshwater sources reduced by 30%.
E	Same as A, except that conveyance from Balqa to Amman has a substantially increased capacity.[a]
F	Same as A, except that Jordan is assumed to subsidize water for agriculture in Irbid and Balqa by charging farmers only 1 cent per cubic meter for up to 92 cubic meters in Irbid and up to 124 cubic meters in Balqa (the amounts used in those districts, respectively, in 1995).
G	Same as E, except that Jordan is assumed to subsidize water for agriculture in Irbid and Balqa by charging farmers only 1 cent per cubic meter for up to 92 cubic meters in Irbid and up to 124 cubic meters in Balqa (the amounts used in those districts, respectively, in 1995).

[a] The capacity of the King Abdullah Canal is left unchanged, however, and in 2020 runs involving subsidized water for Jordan Valley agriculture, this is a binding constraint.

Gaza and wastewater treatment

As already observed more than once, it is always efficient to have a wastewater treatment plant in Gaza and sell the treated wastewater to the Negev for use in agriculture. In 2010, Gaza receives water from the Israeli National Carrier above the present level of 5 million cubic meters (mcm) per year (which WAS takes as the minimum amount) only in the presence of such a plant (see runs A and B). This reflects the desirability of producing the treated wastewater involved. By 2020, that minimum has been greatly exceeded; except in the drought runs (D), conveyance from the Israeli National Carrier to Gaza is well above the 5 mcm. (The existence of the Gaza–Negev reclaimed water system does make a substantial difference, however—about 10 mcm.)

Water from the Jordan River

In 2010, except in drought (run D), Palestine takes water from the Jordan River at Jenin. By 2020, this is no longer the case (except for a small amount in run B). By contrast, except when subsidizing Jordan Valley agriculture (runs F and G), Jordan takes less water from the river in 2010 than the 55

Table 8.2.2A. Summary of Cooperative Run Results with Israeli Fixed-Price Policies, 2010

	A	B	C	D	E	F	G
Jordan River water taken at Jenin (mcm)	47.1	47.2	46.2	0	47.1[b]	8.0[b]	20.0[b]
Freshwater from Israeli National Carrier to Gaza (mcm)[a]	6.0	5.0	5.6	5.0	6.0	5.0	5.6
Treated wastewater from Gaza to Negev (mcm)	11.0	NA	10.8	5.1	11.0	10.5	10.8
Jordan River water taken at Irbid (mcm)	35.0	35.4	NA[c]	24.0	35.0	80.5[d]	81.3[d]
Shadow value at Amman ($/m^3)	0.37	0.37	NA	0.73	0.37	0.39	0.38
Freshwater from Disi to Amman (mcm)	0	0	NA	0	0	0	0
Desalination on Mediterranean coast (mcm)	None	None	None		None	None	None
Acco				80			
Hadera				64			
Raanana				121			
Rehovot				72			
Lachish				29			
Gaza (all districts)				6			
Total				372			

[a] Including 5 mcm already being supplied.

[b] In runs E, F, and G, the variation in the amounts of Jordan River water taken at Jenin is surprising, but illusory. In fact, the total amount of water extracted at Jenin is approximately the same in all three runs. This is because the scarcity rent of Jordan River water in run F (the shadow value of the common-pool constraint on Jordan River water) plus the per cubic meter cost of extraction of that water at Jenin happens to equal the per cubic meter cost of extraction from an alternative source at Jenin. Hence, the optimizing procedure is indifferent as to which source is used. When the scarcity rent of Jordan River water becomes higher (as in the drought run, D), this indifference disappears and Jenin no longer takes water from the river.

[c] In this run, Jordan takes the 55 mcm treaty amount.

[d] With Jordanian agricultural subsidies in effect, the expansion of the conveyance line from Balqa to Amman makes essentially no difference. The shadow value in Amman is not so high as to make water conveyance to that city take precedence over the subsidized Jordan Valley agriculture.

Table 8.2.2B. Summary of Cooperative Run Results without Israeli Fixed-Price Policies, 2010

	A	B	C	D	E	F	G
Jordan River water taken at Jenin (mcm)	50.3	50.7	50.7	0	50.3	48.7	48.7
Freshwater from Israeli National Carrier to Gaza (mcm)[a]	7.4	5.0	7.5	5.0	7.4	6.7	6.7
Treated wastewater from Gaza to Negev (mcm)	11.6	NA	11.7	5.1	11.6	11.3	11.3
Jordan River water taken at Irbid (mcm)	43.6	44.5	NA[b]	32.9	43.6	82.7[c]	82.7[c]
Shadow value at Amman ($/m^3)	0.36	0.35	NA	0.66	0.36	0.36	0.36
Freshwater from Disi to Amman (mcm)	0	0	NA	0	0	0	0
Desalination on Mediterranean coast (mcm)	None	None	None		None	None	None
Acco				72			
Lachish				8			
Gaza (all districts)				6			
Total				86			

[a] Including 5 mcm already being supplied.

[b] Note that in this run, Jordan takes the 55 mcm treaty amount.

[c] Note that with Jordanian agricultural subsidies in effect, the expansion of the conveyance line from Balqa to Amman makes essentially no difference. The shadow value in Amman is not so high as to make water conveyance to that city take precedence over the subsidized Jordan Valley agriculture.

mcm it now receives by treaty.[2] By 2020, however, Jordan takes considerably more water from the river than in 2010. We examine this further when studying Israel–Jordan cooperation below.

Desalination

Desalination on the Mediterranean coast is never required in 2010 except in drought (run D). When it is required, the amount desalinated is much increased by the presence of Israeli fixed-price policies.[3] In 2020, the quantity desalinated under drought conditions is considerably larger—as we should expect, given the larger populations expected.

A new phenomenon arises in the 2020 results, however. Even in times of normal hydrology, some desalination is desirable (at Acco) when both Israel and Jordan subsidize water for agriculture (runs F and G). Of course, these are the runs in which both Israeli and Jordanian agricultural water

Table 8.2.3A. Summary of Cooperative Run Results with Israeli Fixed-Price Policies, 2020

	A	B	C	D	E	F	G
Jordan River water taken at Jenin (mcm)	0	0	0	0	0	0	0
Freshwater from Israeli National Carrier to Gaza (mcm) [a]	20.6	10.4	23.0	5.0	12.3	11.6	11.6
Treated wastewater from Gaza to Negev (mcm)	17.8	NA	18.9	14.2	15.6	15.4	15.4
Jordan River water taken at Irbid (mcm)	52.7	56.0	NA [b]	87.2	77.9	122.8	135.9
Shadow value at Amman ($/m^3)	1.08	1.08	NA	1.14	0.58	1.08	1.08
Freshwater from Disi to Amman (mcm)	17.9	17.9	NA	41.6	0	17.9	3.8
Desalination on Mediterranean coast (mcm)	None	None	None		None		
Acco				93		40.6	53.7
Hadera				67			
Raanana				211			
Rehovot				94			
Lachish				29			
Gaza (all districts)				26			
Total				520			

[a] Including 5 mcm already being so supplied.

[b] In this run, Jordan takes the 55 mcm treaty amount.

demand is highest and Jordan takes considerable amounts of water from the Jordan River.

That is an expected result. As time goes on and water demands grow, with the given water resources and assumed infrastructure, the sensible thing to do under cooperation would be for Jordan to take more water from the river and for Israel to replace some or all of that water through desalination. This does not mean a loss for Israel (or for Jordan), but rather a gain for both parties. We examine the gains to both parties below.

Water for Amman

In 2010, no extreme water crisis appears in Amman,[4] and the line from the Disi Aquifer is not used, consistent with our results in Chapter 7. By 2020, however, the story is quite different. For those years, in all but one of the circumstances applicable to this issue, there is substantial scarcity in

Table 8.2.3B. Summary of Cooperative Run Results without Israeli Fixed-Price Policies, 2020

	A	B	C	D	E	F	G
Jordan River water taken at Jenin (mcm)	5.6	7.8	46.7	0	0	0	0
Freshwater from Israeli National Carrier to Gaza (mcm)[a]	29.3	19.1	30.2	5.0	25.9	27.6	25.9
Treated wastewater from Gaza to Negev (mcm)	21.8	NA	22.2	14.2	20.2	21.0	20.2
Jordan River water taken at Irbid (mcm)	84.2	85.9	NA[b]	93.9	142.8	136.7	143.1
Shadow value at Amman ($/m^3)	1.08	1.08	NA	1.14	0.45	1.08	1.08
Freshwater from Disi to Amman (mcm)	17.9	17.9	NA	41.6	0	17.9	3.8
Desalination on Mediterranean coast (mcm)	None	None	None		None	None	None
Acco				96			
Hadera				2			
Raanana				52			
Rehovot				57			
Lachish				10			
Gaza (all districts)				26			
Total[c]				242			

[a] Including 5 mcm already being supplied.

[b] In this run, Jordan takes the 55 mcm treaty amount, considerably less than is taken in Run A. As a result, Palestine takes more water at Jenin than in Run A. This phenomenon did not occur in Table 8.2.3A, where, with Israeli fixed-price policies in effect, Jordan takes about the same amount of water in Runs A and C.

[c] Separate figures do not add to total because of rounding.

Amman, and the line from Disi is used—and used particularly heavily in drought (run D).

The exception occurs when conveyance capacity from Balqa to Amman is assumed to be expanded even beyond the 90 mcm per year already being built *and* when Jordan Valley agriculture does not receive subsidized water (run E). In these 2020 runs, Amman receives considerable water from Balqa (130 mcm when there are Israeli fixed-price policies and 169 mcm when there are not). That water is sufficient to lower the shadow value in Amman well below the point ($1.08 per cubic meter) at which it is efficient to bring water from Disi.[5]

The situation is different, however, when Jordan subsidizes water for Jordan Valley agriculture. At the assumed subsidized price (1 cent per cubic meter) and the amounts used in 1995, the implied value of water for Jordan Valley agriculture is sufficiently large to cut the transfer of water from Balqa to Amman to 104 mcm (with and without Israeli fixed-price policies). That amount of water is insufficient to bring the shadow value in Amman below the point at which it is efficient to convey water from the Disi Aquifer (it remains at $1.08/m^3$), and a small amount of such water (3.8 mcm) is so conveyed.

Policy Trade-Offs: The Red Sea–Dead Sea Canal

Notice the implications of the preceding point. First, our results suggest that the future water crisis in Amman could be alleviated by increased water from the Jordan River (available under cooperation) if the subsidy to Jordan River agriculture is reduced or eliminated.

Second, if this is not done, then as discussed in Chapter 7, it would indeed be effective, were the Red Sea–Dead Sea Canal to be constructed for other reasons, to supply Amman with desalinated water. In such a case, conveyance from the Disi Aquifer would no longer be efficient.

Evidently, the choices open to Jordan would be considerably wider under cooperation than without it—a conclusion that must also apply in some degree to Israel and Palestine. We now consider the gains that cooperation could bring.

3. Gains from Cooperation: Israel and Palestine

Cooperation using water permit sales creates a win–win situation for all participants. To illustrate this, we begin by examining the gains to Israel and Palestine from such cooperation, starting from varying assumptions about the ownership of the Mountain Aquifer (see Figure P.1). To simplify matters, the case we examine is one in which Israel owns all but 55 mcm of the water of the Jordan River, and the remainder goes to Jordan, as it does at present. This is to be taken as merely an assumption made for the purposes of generating illustrative examples; it is *not* a political statement as to the desirable outcome of negotiations.

We find such gains generally to exceed the value of changes in such ownership that reflect reasonable differences in negotiating positions. Figures 8.3.1–8.3.6 illustrate such findings and more for years of normal hydrology. In these figures, we have arbitrarily varied the fraction of Mountain Aquifer water owned by each of the parties from 80% to 20%. (The present division of the water is about 76% to Israel and 24% to Palestine.[6] Results for that division can be approximated by interpolation but are, of course, fairly close to those for 80% Israeli ownership.)

The two line graphs in Figure 8.3.1A show the gains from cooperation in 2010 for Israel and Palestine, respectively, as functions of ownership allocations. (Throughout, results for Israel are light gray and those for Palestine dark gray.) Israeli fixed-price policies for water are assumed to be the same as in 1995, with large subsidies for agriculture and much higher prices for households and industry.

Starting at the left, we find that Palestine benefits from cooperation by about $170 million per year when it owns only 20% of the aquifer.[7] In the same situation, Israel benefits by about $12 million per year. As Palestinian ownership increases (and Israeli ownership correspondingly decreases), the gains from cooperation fall at first and then rise. At the other extreme (80% Palestinian ownership), Palestine gains about $84 million per year from cooperation, and Israel gains about $36 million per year. In the middle of the figure, total joint gains are about $84 million to $95 million per year.[8]

It is important to emphasize what these figures mean. As opposed to autarky, each party benefits as a buyer by acquiring cheaper water. Moreover, each party benefits as a seller over and above any amounts required to compensate its people for increased water expenses.

Why do the gains first decrease and then increase as Palestinian ownership increases? At the extremes, large gains can be made by transferring water from the large owner to the other party. Palestine has larger benefits at the left-hand side of the diagram than in the middle because it can obtain badly needed water from cooperation; it has larger gains at the right-hand side than in the middle because, when it owns most of the Mountain Aquifer water, it can gain by selling relatively little-needed water to Israel (which gains as well). The same phenomenon holds in reverse for Israel, although here the effects are smaller, largely because Israel is assumed to own a great deal of water from the Jordan River.

One might suppose that the gains would be zero at some intermediate point, but that is not the case. It is true that a detailed, noncooperative water agreement could temporarily reduce the gains from cooperation to zero. That would require that the agreement exactly match in its water-*ownership* allocations the optimizing water-*use* allocations of the optimizing cooperative solution. That is very unlikely to happen in practice (and if it did, this optimal solution would not last as populations and other factors changed). In our runs, it does not happen for two reasons.

First, we have not attempted to allocate ownership of the Mountain Aquifer in a way so detailed as to match geographic demands. Instead, we have allocated each common pool in the aquifer by the same percentage split.

Second, there are gains from cooperation in these runs that do not depend on the allocation of the Mountain Aquifer. For example, it is always

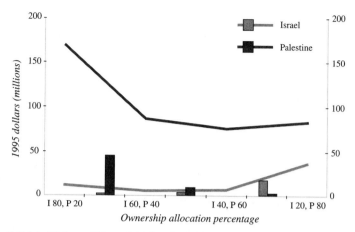

Figure 8.3.1A. Value of Israel–Palestine Cooperation and Value of Ownership of Mountain Aquifer without Cooperation and with Israeli Fixed-Price Policies, 2010

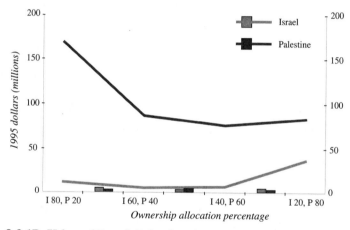

Figure 8.3.1B. Value of Israel–Palestine Cooperation and Value of Ownership of Mountain Aquifer with Cooperation and with Israeli Fixed-Price Policies, 2010

efficient for treated wastewater to be exported from Gaza to the Negev for use in agriculture.

There are further results to be read from Figure 8.3.1A. The height of the bars shows the value to the parties *without cooperation* of a change in ownership of 10% of the Mountain Aquifer (about 65 mcm per year, or nearly half the amount of Mountain Aquifer water now taken by Palestine). These are shown as functions of ownership positions midway within each 20-percentage-point interval. For example, the bars on the left show the value to each of the parties of an ownership shift of 10% of the Mountain

Aquifer, starting at an allocation of 70% to Israel and 30% to Palestine; the center bars indicate the value of a such a change starting at 50–50. Note that the value of cooperation is generally greater than, or at least comparable to, the value of such ownership changes. This is especially true for Palestine, but the result holds for Israel as well.

Now look at Figure 8.3.1B, which differs from Figure 8.3.1A only in the height of the ownership-value bars. The height of these bars represents the value of shifts of 10% aquifer ownership *with cooperation*. That value is about $8 million per year.

The lesson is clear: ownership is surely a symbolically important issue, and symbols really matter, but cooperation in water reduces the practical importance of ownership allocations—already not very big—to an issue of very minor proportions.

The same qualitative results hold when we examine Figures 8.3.2A and 8.3.2B. These differ from Figures 8.3.1A and 8.3.1B, respectively, in that Israeli fixed-price policies, which subsidize agriculture but charge prices above shadow values to industry and households, are assumed not to be in effect and water is sold to users at the efficiency prices generated by WAS.

There are three differences between the two pairs of figures that are worth discussing.

- Without cooperation, although the value to Palestine of ownership changes remains the same (as they must, since Palestine receives exactly the same water as before), the gains to Israel are reduced, but reduced significantly only when Palestine owns the lion's share of the aquifer.
- With cooperation, the value of ownership changes is even a bit smaller in Figure 8.3.2B than in Figure 8.3.2A.
- The gains from cooperation are not much different.

One of the issues that might arise in contemplating a cooperative agreement of the type described is as follows. If Israel subsidizes water for agriculture, then there will be the following two effects on Palestine (and a similar analysis applies to subsidies by Palestine and effects on Israel).

Output effect: An agricultural water subsidy reduces the cost of production borne by Israeli farmers. This enables sales at lower prices than would be possible without the subsidy, reducing the demand for Palestinian agricultural output.

Water effect: An agricultural water subsidy increases the demand for water by Israeli farmers. Since this increases water scarcity, it will increase the shadow values of water and hence the prices to Palestinian consumers (or to their government, if the government chooses to fix prices to consumers).

Does not this mean that a cooperative agreement will lead to continual negotiation over Israel's price policy? So far as the output effect is concerned, this is not really a water issue. The same effect would occur (and

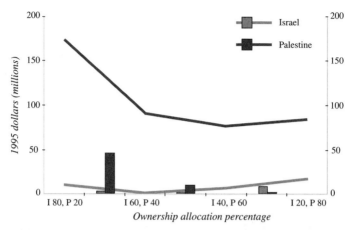

Figure 8.3.2A. Value of Israel–Palestine Cooperation and Value of Ownership of Mountain Aquifer without Cooperation and without Israeli Fixed-Price Policies, 2010

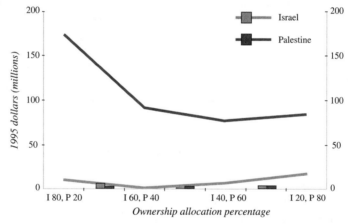

Figure 8.3.2B. Value of Israel–Palestine Cooperation and Value of Ownership of Mountain Aquifer with Cooperation and without Israeli Fixed-Price Policies, 2010

occur more efficiently) if Israel simply subsidized agriculture directly without interfering with the choice of inputs. Hence, this is not really an issue that comes about because of a cooperative agreement in water. That is not true of the water effect. However, perhaps surprisingly, that effect turns out to be at worst quite small, and at best negative—a benefit to Palestine.

Figure 8.3.3 shows the difference to each party made by Israeli fixed-price policies in the context of a cooperative agreement. The negative effects on Palestine are very small at worst (so that negotiations over Israel's fixed-price policies should not be a major issue). They are about $4 million with Israeli ownership of 80% of the aquifer, and indeed, with increases in

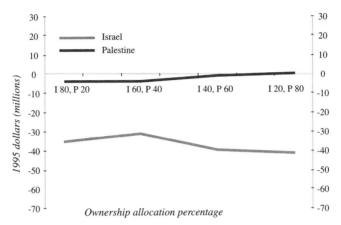

Figure 8.3.3. Effect of Israeli Fixed-Price Policies on Total Surplus for Israel and Palestine under Cooperation, 2010

Palestinian ownership, the effects rise toward zero, eventually even becoming slightly positive. This occurs because fixed-price policies also increase the price that Israel must pay to obtain Palestinian water, and with increasing Palestinian ownership, the amount of such purchases rises.

Of course, the corresponding effect on Israel itself is in the other direction. The effect of fixed-price policies on Israel starts off negative and remains so, eventually increasing in size. It must be remembered, however, that this is the price of having such policies, particularly of subsidizing agriculture. Presumably, Israel's policymakers consider that there is an added social gain from doing so—a gain not reflected in the calculations shown.

Figures 8.3.4–8.3.6 show similar results for 2020. As we should expect, all the monetary figures are greater (for example, total gains from cooperation range from $116 million to $232 million per year, instead of $84 million to $184 million), but the qualitative conclusions are the same. Note in particular that the value of an ownership shift of 10% of the Mountain Aquifer under cooperation is only about $3.5 million to $11.5 million per year. Although water will be more valuable in 2020 than in 2010, mostly because of population growth, the value of cooperation will still generally be higher for each party than a gain of ownership of 10% of the aquifer. (As in all such figures, the comparison must be made between a line of one shade of gray and bars of the same shade of gray.)

The water effect of Israeli fixed-price policies on Palestine remains small relative to Palestinian gains from cooperation and—save where Israel owns most of the aquifer—absolutely small, becoming definitely positive where Palestine owns the major part.

As discussed in Chapter 4, however, the greatest benefits from cooperation may not be monetary. Beyond pure economics, the parties to a water agreement would have much to gain from an arrangement of trade in water

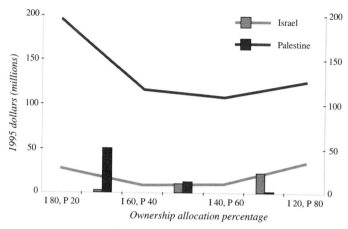

Figure 8.3.4A. Value of Israel–Palestine Cooperation and Value of Ownership of Mountain Aquifer without Cooperation and with Israeli Fixed-Price Policies, 2020

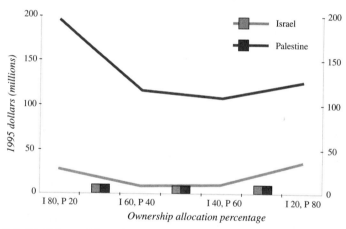

Figure 8.3.4B. Value of Israel–Palestine Cooperation and Value of Ownership of Mountain Aquifer with Cooperation and with Israeli Fixed-Price Policies, 2020

permits. Water quantity allocations that appear adequate at one time may not be so at other times. As populations and economies grow and change, fixed water quantities can become woefully inappropriate and, if not properly readjusted, can produce hardship. A system of voluntary trade in water permits would be a mechanism for flexibly adjusting water allocations to the benefit of all parties and thereby for avoiding the potentially destabilizing effect of a fixed water quantity arrangement on a peace agreement. It is not optimal for any party to bind itself to an arrangement whereby it can neither buy nor sell permits to use water.

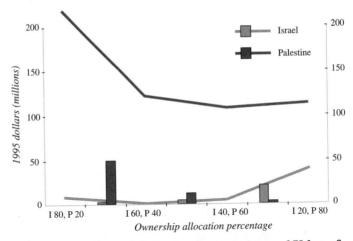

Figure 8.3.5A. Value of Israel–Palestine Cooperation and Value of Ownership of Mountain Aquifer without Cooperation and without Israeli Fixed-Price Policies, 2020

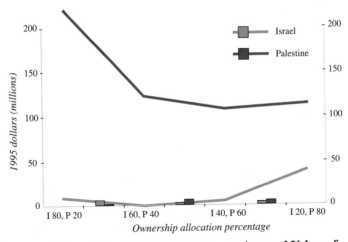

Figure 8.3.5B. Value of Israel–Palestine Cooperation and Value of Ownership of Mountain Aquifer with Cooperation and without Israeli Fixed-Price Policies, 2020

Moreover, cooperation in water can assist in bringing about cooperation elsewhere. For example, as already indicated, the WAS model supports the idea of building a wastewater treatment plant in Gaza and selling treated effluent to Israel for agricultural use in the Negev. Both parties would gain from such an arrangement. Thus Israel has an economic interest in assisting with the construction of a Gazan wastewater treatment plant—a serious act of cooperation and a confidence-building measure that would not impinge on the core values of either side.

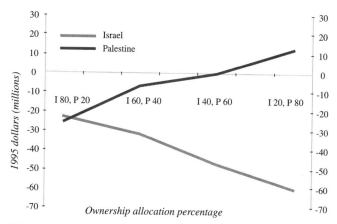

Figure 8.3.6. Effect of Israeli Fixed-Price Policies on Total Surplus for Israel and Palestine under Cooperation, 2020

4. Gains from Cooperation: Jordan and Israel

We now consider examples involving Jordan, starting with that of cooperation between Jordan and Israel (adding Palestine below). Here we examine different divisions of ownership of the Jordan River between Israel and Jordan. For convenience of illustration—and again without intending any political statement—we here assume that Israel owns 76% of the Mountain Aquifer (roughly the amount it now uses) and that neither Israel or Jordan cooperates with Palestine.

We present fewer results than for Israel and Palestine because we assume in the figures below that both Israel and Jordan maintain fixed-price policies, relaxing that assumption only briefly later on.

Figures 8.4.1A and 8.4.1B show the results for 2010 and Figures 8.4.2A and 8.4.2B for 2020. Jordan is represented in dark gray and Israel again in light gray. The interpretation of the lines in the graph is again the gain from cooperation; that of the bars is again the value of an ownership shift of 10% of the resource (about 72 mcm per year). We start the lines at the existing division of the water (662 mcm taken by Israel and 55 mcm by Jordan[9]) but start the bars only at the 80–20 division point, then move on by 20% increments, as we did in the previous section. (Note that the horizontal axis is not intended to be quantitative.)

Again, we see that the gains from cooperation are bigger than those from shifts in ownership. In particular, when Jordan owns much of the river, it gains essentially nothing from further ownership without cooperation, but it gains very substantially by selling water permits to Israel under cooperation. Correspondingly, Israel gains by buying water permits.

In general, gains from cooperation are positive at the existing division—especially so for Jordan in 2020—but fall to approximately zero when Israel

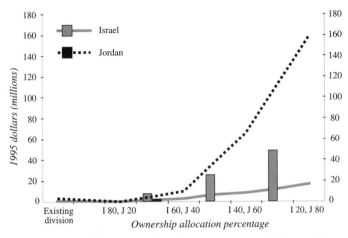

Figure 8.4.1A. Value of Israel–Jordan Cooperation and Value of Ownership of Jordan River without Cooperation and with Fixed-Price Policies, 2010

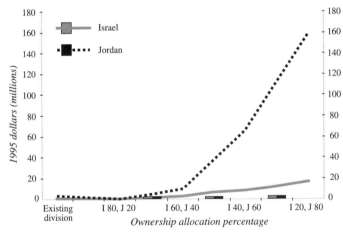

Figure 8.4.1B. Value of Israel–Jordan Cooperation and Value of Ownership of Jordan River with Cooperation and with Fixed-Price Policies, 2010

owns 80% of the water, and then rise substantially as Israel begins to buy water permits from Jordan.[10]

The value of 10% ownership shifts under cooperation is again very low (about $3.5 million per year in 2010 and about twice as much in 2020—corresponding to only about $10 million per year for 100 mcm per year). Further, a new phenomenon can be observed here. As opposed to the case of the Mountain Aquifer (Figures 8.3.1B and 8.3.4B), the value of ownership shifts is constant (except for rounding error). This reflects the fact that the Jordan River is a single resource, so ownership transfers under cooperation correspond to changes in expenditures and receipts from sales of water

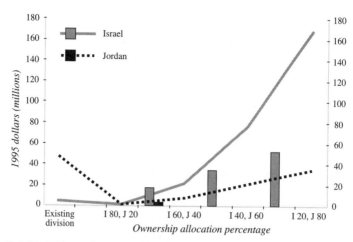

Figure 8.4.2A. Value of Israel–Jordan Cooperation and Value of Ownership of Jordan River without Cooperation and with Fixed-Price Policies, 2020

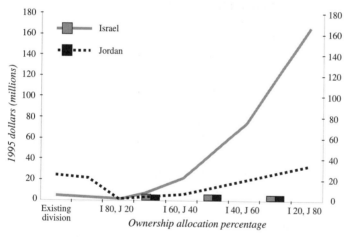

Figure 8.4.2B. Value of Israel–Jordan Cooperation and Value of Ownership of Jordan River with Cooperation and with Fixed-Price Policies, 2020

permits at the same scarcity price. The Mountain Aquifer, on the other hand, consists of several sources. Changes of ownership under cooperation in that case also correspond to changes in expenditures and receipts from sales of water permits; those sales, however, involve several different scarcity prices, and the average price involved can differ depending on the starting ownership position's influence on how much water from each source is involved.

The difference between allocating a single source and allocating multiple sources is also the reason that gains from cooperation fall to zero for a particular water allocation here but did not do so in the case of the Mountain Aquifer. But even with such an allocation made, gains from cooperation will appear as circumstances change. One can never do better without cooperation than one can with it; one can only temporarily do as well.

We have also examined the water effect on Israel of Jordanian agricultural subsidies. At the existing division of Jordan River water, Israel gains slightly ($1 million per year in 2010 and $3 million per year in 2020) from the presence of such subsidies.

5. Gains from Cooperation: Adding a Third Partner

We now consider the effect on cooperative gains of adding a third partner to an existing bilateral cooperative arrangement. We do so by adding Palestine to an Israel–Jordan arrangement and by adding Jordan to an Israel–Palestine arrangement. In both cases, we work with a 76–24 division of the Mountain Aquifer (approximating the existing circumstances) and the existing division of the Jordan River.

Table 8.5.1 presents the results for 2010 and 2020. In both cases, we assume that Israel and Jordan have fixed-price policies, since that is probably the leading case, and the phenomenon found can easily be seen to generalize.

In all cases, Israel gains and its existing partner loses when the third party enters the arrangement.[11] However, Israel gains more than the existing partner loses, and of course, the entering party also gains. Hence, it is always possible to compensate the losing partner for permitting the arrangement to expand, leaving all three parties with a net gain.

It is easy to see that this result is perfectly general (and generalizes to the case of more than three parties as well). Consider a cooperative water permit arrangement among N parties. Call the set of those N parties S. Since the parties in S behave cooperatively, they effectively act as a single unit. Hence the addition of an $N+1$st party must bring a nonnegative gain both to the additional party and to S considered as a unit, since that is true of any bilateral cooperative arrangement. But then it must be possible to redistribute the gains, if necessary, so that everybody wins.

In the cases examined in Table 8.5.1, Israel always gains and the existing partner always loses (before redistribution). This is not a general property, and indeed, there can be cases in which all three parties gain without redistribution.

The reason for the phenomenon in Table 8.5.1 is easy to find. With the assumed ownership divisions, Israel has a great deal of water and acts as a net water permit seller in its cooperative arrangements. Palestine and Jordan, on the other hand, are net water permit buyers. The entry of either into an arrangement between Israel and the other necessarily increases the demand for water permits from Israel and raises the price at which such

Table 8.5.1. Effects of Adding a Third Partner to Bilateral Cooperation ($million per year)

Change	Effects on existing partners	Gain to new partner
2010		
Addition of Jordan to Israel–Palestine cooperation	Israel gains $2[a] Palestine loses $2[a]	Jordan gains $2
Addition of Palestine to Israel–Jordan cooperation	Israel gains $10 Jordan loses $2	Palestine gains $119
2020		
Addition of Jordan to Israel–Palestine cooperation	Israel gains $8[a] Palestine loses $9[a]	Jordan gains $31
Addition of Palestine to Israel–Jordan cooperation	Israel gains $28 Jordan loses $15	Palestine gains $114

[a] The appearance of zero or slightly negative net gains to the existing partners taken together is due to rounding error. See text.

permits get sold (by increasing the scarcity rent of the water involved). This benefits the net seller but disadvantages the net buyer. (Of course, if a new net seller were to enter, the situation would be reversed.) Fortunately, it is always possible to redistribute the gains so that everybody benefits from increased cooperation.

6. Conclusion: Water Values, Economic and Noneconomic

Finally, we naturally come to a question to which we have so far given relatively little attention. This book has been about the economics of water. Although "economics" can (and should) be broadly construed to permit social values to enter that are not merely private values, we have concentrated on the value of water considered as molecules of H_2O. But water in certain sources—the Jordan River in particular—can have important historical, religious, symbolic, or emotional value. What can we say about that?

We cannot evaluate such values, but we can describe what happens if a party treats a certain body of water as special and withholds it from a cooperative water permit agreement. Suppose, for example, that Israel were to reserve the water of the Jordan River, participating in cooperation only as regards its other water sources. Such an action would naturally disadvantage the other parties to a water permit cooperative agreement. But it would also be costly to Israel itself. Just as refusal to trade water permits generally gives up benefits that would otherwise arise from such trade, so too does a refusal to include specific water sources in such an agreement.

How much would it cost the refuser? WAS can be used to answer that question, but in the case of Israel and the Jordan River, the answer must be very dependent on the ownership assumptions made, and we have not investigated this question quantitatively. Should the day ever come, however, when cooperation is in sight, the answers to such questions of practical importance can readily be obtained. In this particular case, Israel would have to decide whether the special nonmonetary benefits of withholding the Jordan River from an agreement are worth the costs incurred to do so.

A judgment that it is worth the costs is entirely possible. As do all countries, the three parties studied here have values that are not easily reduced to money values. Further, they unfortunately have many matters over which they disagree—sometimes, alas, violently. But water *as water* should not be one of those. We believe we have shown that water as water should not become an obstacle to agreement either in the Middle East or anywhere else on earth.

Notes

Preface

1. Reuters interview, reported on Environmental News Network, August 13, 2001.
2. Paul Simon, "In an Empty Cup, a Threat to Peace," *New York Times,* August 14, 2001.
3. For example, see Klare (2001, 56–57, 59–60).
4. We use the term "Palestine" throughout without intending a particular view as to the ultimate resolution of the Israeli–Palestinian conflict, even though each of the coauthors doubtless has such a view. It would be disrespectful to our Palestinian colleagues to do otherwise, and writing so as to avoid the term would be immensely awkward.
5. The map in Figure I.1 calls the large lake on the Jordan River the Sea of Galilee. That lake is called the Kinneret by Israel and Lake Tiberias by Jordan and Palestine. All of these are correct. Hence, in the chapter dealing with Israel (Chapter 5), we use "the Kinneret," and in the chapters dealing with Palestine and Jordan (Chapters 6 and 7, respectively), we use "Lake Tiberias." In chapters involving more than one party, we use the relatively neutral name "Sea of Galilee."
6. It is important to note that what is being discussed in the text is the valuation of water *as water.* Of course, particular water sources may be valued for historical or ideological reasons. But these really have to do with attachment to the land, not to molecules of H_2O. This book is about the latter.
7. This assumes that the efficient use of such water would be to supply the coastal cities. That is almost certainly true, but even if it were not, the principle being illustrated would still hold (with different numbers).
8. There is no AGSM section here. The Palestinian treatment of AGSM in the project did not produce striking results but was very useful for its insights into the issues that can be encountered in applying AGSM. That discussion has been incorporated in Chapter 3.
9. This history mentions a number of people, but these are by no means all who contributed. A more complete acknowledgments section is given at the end of this preface.

10. This was later published in Eckstein et al. (1994).
11. This statement relates to prices per cubic meter. As discussed in later chapters (especially Chapter 7), it does not hold in such a simple way when conveyance links are used to capacity.
12. When the Middle East Water Project was adopted by the initiative in January 1996, Egypt decided to play no active role but to act as an observer.
13. Israel had no planning ministry as such. Beilin was then economics minister.
14. Shamir and especially Arlosoroff had provided considerable advice and counsel during Phase 0. The same was true of Munther Haddadin in Jordan.
15. There may have been several reasons for this. The reason given was that the Dutch were refusing to fund the project in a way that would finance the new administration of the Institute.
16. For examples that go beyond mere cost minimization, see Brown and McGuire (1967), Dandy et al. (1984), and McCarl (1999). The model that appears most similar to WAS is the CALVIN model, an optimizing water model for California developed at the University of California, Davis. (See, e.g., Newlin et al. 2002; Jenkins et al. 2003, 2004.)
17. We particularly apologize for any omission of names that should be listed in this section and any misidentification of roles. Harvard has refused access to the project's files, and we can no longer recover with certainty all the names and roles of the persons involved.

Chapter 1

1. Microeconomics is the study of prices, households, firms, markets, and the allocation of resources. Macroeconomics (which is what most noneconomists think of when they think of economics) is the study of large aggregates, such as gross domestic product, unemployment, inflation, and so forth.
2. It was this observation by the late Gideon Fishelson that set our project in motion (see the Preface). It must be understood that when we speak of the value of water, we are speaking of the value of molecules of H_2O—of water as water. The value that attaches to particular bodies of water for historical or religious reasons is not the same thing.
3. This is an illustration of the Coase Theorem of economics; see Coase (1960).
4. For simplicity, we assume that Ozite, once extracted, can be used only once, so secondary Ozite need not be considered.
5. Maximizing present value is the generalization of maximizing profits when more than a single period is considered.
6. This discussion assumes that Ozite can be stored only by leaving it in the ground or, at least, that below-ground storage is cheaper than above-ground storage. Dropping this assumption would merely complicate the exposition without adding anything worthwhile.
7. The proposition that private mine owners will maximize present value by acting so that marginal profit increases at the interest rate is known as Hotelling's Rule; see Hotelling (1931).
8. The issue of coincidence of private and social discount rates is also of importance, but discussion of it is not needed for later purposes.
9. Note that this could also be treated as a case in which there are social costs to solely private water use that are not reflected in private costs.

10. For a more detailed description of what follows, see Rostow (1948).

11. The particular legal form does not matter. We have chosen a corporation for simplicity.

12. This is closely related to the Coase Theorem, cited above.

13. For households, this is an approximation, although it is likely to be a good one. For industry and agriculture, the calculation is exact.

14. These are LaGrange multipliers. We use the term *shadow values* rather than the more usual *shadow prices* to avoid confusion with the prices that water users actually pay. The latter prices may but need not be shadow values.

15. If this calculation gives a negative figure, then scarcity rent is zero, and water is not scarce at the given location.

16. This requires the same Jacobian condition as before.

Chapter 2

1. This would also imply that the area under the demand curve is infinite. WAS deals with the latter problem by taking the lower limit of integration over quantity not as zero but as the quantity at which equation 2.1.1 would imply a price of $100 per cubic meter. So long as one is careful not to attribute meaning to the level of total benefits but only to changes or differences in that value in different runs of the model, this presents no difficulty.

2. These subscripts should not be confused with that used above in P_0 and Q_0 to denote particular values of P and Q, respectively.

3. This fact is one of the basic reasons that private water markets will not solve the water allocation problem. See Chapter 1. It can also be true that extraction at one source will affect extraction costs at other sources. Some investigation of this issue was done by Y. Bachmat in connection with Phase 0 of our project but has not been included in the current version of WAS. The effects involved appear relatively minor.

4. Note that this must be done even when the different districts belong to the same political entity. Obviously, when the underlying water source is claimed by more than one such entity, the problem is even more important, although not technically different from the single-entity case.

5. Capacity expansion should not be thought of here as literally adding one unit to the capacity of an existing facility. Rather, what is involved is planning in advance to make an unbuilt facility larger by one unit. There is a general theorem by Harold Hotelling that shows that optimal pricing would have the present value of all future capacity shadow values appear in the price and also add up to the cost of a unit of capacity. That theorem was shown to us by the late Robert Dorfman nearly a decade ago. Despite considerable effort, we have been unable to recover the reference. (See Footnote 17 of the Preface.)

6. In the simplest version, the stream is constant. If changes in demand conditions are expected, however (for example, because of population growth), then runs can be made for selected future years to estimate a nonconstant benefit stream.

7. We say "would be indicated" because for some users, the price paid if no policy is imposed is not the shadow value itself but rather the shadow value adjusted for effluent charges and wastewater treatment profits. See Section 3.

8. The notation used is restricted to this appendix.

9. Note that the first term of the objective function is the integral of the *inverse* demand function: $P_{id} = B_{id} \times (QD_{id} + QFRY_{id})^{\alpha_{id}}$. Aside from the subscripts and the fact that water quantity is given as the sum of two parts, this corresponds to the demand function equation 2.1.1 after it is solved for price. In particular, in terms of equation 2.1.1, $\alpha = 1/\eta$ and $B = A^{-1/\eta}$.

10. See Figures 2.A.1A and 2.A.1B for an illustration of continuity for freshwater and recycled water, respectively, as given in the first two equations listed below.

Chapter 3

1. One dunam = one tenth of a hectare = 1000 square meters.

2. These correction factors can be an important contribution to model formulation because they enable the user to include in a large-scale model specific data related to smaller scales.

3. Results for Palestine were also obtained but do not appear to offer sufficiently interesting features to warrant separate presentation. The very insightful comments of Ammar Jarrar, who was responsible for that work, have been incorporated into the general discussion.

4. Elasticity of demand is defined as $(\partial \log Q / \partial \log P)$, where Q and P denote quantity and price, respectively. This is approximately the percentage change in quantity demanded that results from a 1% increase in price. For a linear demand curve, elasticity is not constant, so we report it at a representative price.

5. This suggestion was first made by Annette Huber-Lee. Note that in the case of a single water type, the agricultural demand curve for water simply graphs the marginal revenue product of water (the derivative of WRC with respect to water quantity), and if we take the integral under the agricultural demand for water, we obtain WRC. Hence putting WRC into the objective function is equivalent to putting in that integral. The same proposition holds when there are many water types, only here the use of WRC is far simpler than the calculations necessary to use the multidimensional equivalent of the integral under the demand curve.

Chapter 4

1. We write in terms of international disputes, but of course, the same analysis applies to disputes between states or provinces or cities within a country and to disputes among different water-using groups.

2. If water and money are equally valued, then the entity will be indifferent between selling and using the amount of water in question. A similar statement applies to the description of the actions of a nonowning buyer later in the paragraph. Note that the statements made in the text do not assume that water users value water only for purely economic reasons; they assume that users can consistently choose between water and money.

3. This is an application of the well-known Coase Theorem of economics (Coase 1960).

4. See the Preface for details. All monetary values in this book are in 1995 dollars.

5. It must be emphasized that these results (and those given later) were obtained using data not officially approved by the authorities. Further, not all authors necessarily agree with all the policy prescriptions.

6. In this chapter, we use examples concerning Israel and Syria and Israel and Palestine to illustrate the principles involved. A more complete treatment of results—and one that explicitly includes Jordan—is given in Chapter 8.

7. The costs to Israel given in the text are costs to the country as a whole after actions are taken to reoptimize water allocations. Of course, costs will be greater if the water is simply given up and no offsetting measures are taken.

8. Of course, the quantitative results quoted depend on the data and assumptions made. We are informed that other means of calculation yield different numerical results. The qualitative conclusions, however, would be the same.

9. Trading in such permits occurs in actual water markets in the American West—but not, of course, between the United States and other countries.

10. Although, surprisingly, not because of increases in the Gazan population. See Chapter 6.

11. Alternatively, the Palestinians might seek alternative sources of supply from Egypt or elsewhere—sources that might be efficient even in the presence of trade.

12. Sadly, however, we note that reliance on other parties to act in their own self-interest may prove a weak reed (and this applies to all parties in the region). A well-known story illustrates this point:

> A duck is swimming in the Suez Canal when he is called to the bank by a scorpion.
>
> "Oh, duck," says the scorpion, "I can't swim and I really need to get across the canal. Please put me on your back and take me across."
>
> "You must think I'm stupid," says the duck. "You're a scorpion, and when we get halfway across, you'll sting me."
>
> "Oh, I would never do that," replies the scorpion. "If I were to sting you, you would die. Then you would eventually sink, and I would die, too, because I can't swim, which is why I need you in the first place."
>
> "That makes sense," says the duck. "Hop on."
>
> So the scorpion gets on the duck's back. Halfway across, the scorpion stings the duck. With his dying breath, the duck says, "Scorpion, why?"
>
> "What do you expect?" says the scorpion. "This is the Middle East."

13. In our results, the desalination upper bound considerably overstates the value of such property rights.

Chapter 5

1. All the data and projections presented can be easily changed by the user from the interface of the WAS computer program.

2. The Water Commission uses fewer districts for its planning and data collection. It is possible to change the model to their system, but this would require considerable adjustment and rearrangement of the data.

3. Some adjustment of this assumption is permitted by assuming that either a fixed cost or a fixed benefit accompanies the use of recycled water. In addition, the use of recycled water can be limited by the user of the model. The second of these facilities was used to limit recycled use in the Raanana district to 5.65 mcm per year, the actual quantity used there in 1995.

4. All monetary values in this book are in 1995 dollars.

5. Of course, such constant elasticities cannot apply when prices are very high. We arbitrarily cut off demand at $100 per cubic meter (see Chapter 2).

6. We have not used any classified material.

7. As already discussed, the name given to this body of water—also known as the Sea of Galilee and Lake Tiberias—is that used in the country studied.

8. More accurately, the costs of using such water are not dealt with in the model. Further, the quantities involved are permitted to vary (as do all naturally occurring water supplies) when the user employs the "supply multiplier" facility of the interface to simulate a drought or a wet year.

9. There are other common-pool constraints involving the Mountain Aquifer, but these are not binding for runs involving Israel alone, since the sum of the quantities for the districts and supply steps involved does not exceed the amount of water in the common pools.

10. This differs from the total consumption figure given in Table 5.1.1 because of agricultural use of recycled waste water and other nonpotable water.

11. This is the average nationwide figure for costs beyond the treatment costs necessary for environmentally safe disposal.

12. The sources for these data are the same as those for the supply data and are listed above.

13. We do not give the capacity of the various conveyance links, which are usually quite large. As explained below, we begin by assuming that there are no effective capacity constraints on conveyance so as to make the most favorable case for desalination. An explicit example of the analysis of conveyance capacity and its expansion is provided later in the chapter.

14. It may well be that this is too tight a restriction for later years. However, if so, using it only strengthens our conclusions below.

15. We assume this water to come from the Jordan River (Kinneret), since this is the most efficient way of supplying it, but the WAS model can evaluate the costs of using another source.

16. Fixed-price policies for agriculture involve an increasing three-step block. The first block is set at 17 cents per cubic meter, the second at 20 cents, and the final at 27 cents. These increasing blocks are entered into the model for each district. Actual policies are applied at the farm level. These are aggregated to get the district-level estimates.

17. The apparent discrepancies between the figures in Table 5.2.1 and those in Table 5.1.1, above, are due to the use of recycled wastewater by agriculture, which is not broken out separately in Table 5.1.1.

18. Close observers may wonder why the shadow values for the administrative run do not equal the cost per cubic meter of the most expensive supply step used (plus conveyance costs, if any). The answer is that, as discussed above, the three districts that can extract water from the Jordan River (Hula, Golan, and Biqaat Kinarot) are connected by a common-pool constraint reflecting the fact that they draw on the same water. That constraint has a shadow value of 2 cents per cubic meter, the scarcity rent of Jordan River water in the administrative run. That scarcity rent forms part of the shadow value of Jordan River water wherever that water is used.

19. This term was defined in Chapter 3 as the difference between the revenue received for a crop and the nonwater expenses required to produce that crop, per dunam.

20. The fact that the government may itself, in another capacity, be the water producer is irrelevant here. In that case, we are dividing the government's role into two parts.

21. Because we force the model to send 5 mcm per year to Gaza, the model output shows a positive number for such constrained capacity charges as well as a small international payment from Palestine to Israel for the water. Both payments are the same in both the competitive and the administrative runs, and we ignore them. Of course, when we come to cooperative arrangements in a later chapter, such payments form part of the accounting system.

22. No such treatment is required for capacity-constrained desalination plants because the shadow value of the capacity constraints in such plants is not only reflected in the price paid by buyers (either water consumers or the government) but also counted as received by the producers of the desalinated water.

23. A little thought will reveal that it is irrelevant to this discussion whether the revenues received from consumers exactly cover, more than cover, or fall short of covering the capital costs.

24. As before, only differences in buyer surplus are given.

25. See the appendix to this chapter.

26. The same effect implies an overstatement in the difference in total buyer surplus reported in Table 5.2.2 with an exactly offsetting understatement in the difference in government costs and hence no change in the estimate of deadweight loss.

27. These involved three steps for agriculture, starting at approximately 17 cents per cubic meter and ending at approximately 27 cents per cubic meter. This usually made the marginal cost of water higher than it had been in 1995.

28. In 2004 Israel signed contracts to import water from Turkey at a somewhat higher price. Of course, there are international-relations benefits from such a deal.

29. Note that even with desalination on the coast, this occurs for the Jordan Valley settlements, which would have to be supplied by conveyance in such a case.

30. The model also suggests that desalination would be efficient at Eilat on the Red Sea. The required plant size is 21 mcm per year.

31. When a conveyance line at 10 cents per cubic meter is added from Beit Shean to the Jordan Valley settlements to avoid infeasibility of the fixed-price policies there, the desalination plant in the Raanana district is enlarged by about 9 mcm per year, the same amount that passes over the conveyance line.

32. Note, in this regard, that the infeasibility problem in the Jordan Valley settlements disappears.

33. The capacity required at Eilat, on the other hand, would be less than 1 mcm per year, compared with 21 mcm per year in the presence of fixed-price policies.

34. The frequency calculations in this paragraph are made by assuming the relative frequency to be approximately the same as that with which rainfall in Jerusalem or Nablus falls short of its long-term average by 30% or more. We are indebted to Yoav Kislev for the latter figures.

35. If the Jordan Valley settlements are not partly supplied through conveyance from Beit Shean (necessary to avoid infeasibility of the fixed-price policies there), then the Hadera plant produces about 4 mcm less per year than is shown in the table.

36. As before, somewhat more capacity is required to make the fixed-price policies feasible for the Jordan Valley settlements. The same is true of the 30% case.

37. The Litani is used only as a case study of a new source of water to be imported in the North. Imports from other sources (in Turkey, for example) could also be studied but would naturally have different data and possibly lead to different results. The Litani-Awali discharges an average of more than 500 mcm per year into the Mediterranean. There is no foreseeable economic use for the total quantity. A surplus of 200 mcm per year is very reasonable to assume as tradable under peaceful conditions.
38. We make the minor correction needed to avoid infeasibility in the Jordan Valley settlements.
39. The calculation is rough because the shadow values give only the value of the marginal cubic meter of water. The exact calculation for 100 mcm requires running the model with the amount of available water changed by 100 mcm and calculating the difference in net benefits.
40. Note that all our examples should be treated as such. They depend on numbers that may not reflect actual future facts, but the methods are applicable and powerful.
41. Note that our earlier results on desalination and water imports would only be strengthened if we introduced such constraints. With a lower ability to convey water from the coast, the shadow values of water in coastal districts will be *lower* and additional water there even less valuable than reported above.
42. Since the population in the district is projected to be larger in 2020 than in 2010, this may seem strange. The results occur because in the 2020 runs, it is more efficient to produce somewhat more freshwater from a relatively high-cost source in the Jerusalem Mountains district and import less from the rest of the system, where water shadow values are higher than for 2010. One must not think of only a single district in understanding the results. Naturally, such estimates and the specific numerical results below depend in part on the population forecast for the Jerusalem Mountains district. That population is projected as growing from 639,000 in 1995 to 798,000 in 2010 and 892,000 in 2020.
43. We also note that we are now in a case in which the social benefits accounting discussed above involves payments for the use of constrained pipeline capacity. As explained above, those payments are reflected in the shadow values of the pipeline in Jerusalem but remain in the system. (In the case of fixed-price policies, one can think of such payments as a transfer within the government.)
44. See Chapter 3.
45. The following section is taken from Amir and Fisher (2000). Reprinted with permission from Elsevier.
46. There are, however, two very serious limitations to the selection of unirrigated crops as an alternative to irrigated crops in Israel: winter grains need more than 300 mm rainfall per year, and rainfall in Israel is a stochastic resource in its amounts, place, and time. Therefore growing rain-fed crops is risky. When unirrigated crops are considered as an alternative, the risk involved in rainfall should be reflected in the WRC of unirrigated winter crops. The reduction should be a function of the variance of the statistical distribution of rain in the district under consideration.
47. To simplify discussion, we ignore the fact that there are two water qualities involved. This can be safely done, since the total amount of recycled water used is the same in all runs.
48. This was first pointed out by an anonymous referee of Amir and Fisher (2000).

49. The value of P^* is the same for the two quota amounts because the same crops (orchards) are the marginal crops in the basis for both amounts. In effect, the linear programming nature of the model generates a discontinuous demand curve. But real demand curves are not discontinuous. This problem can be overcome by dropping the assumption that all plots planted to the same crop are identical in WRC and water requirements, as discussed in the appendix to Chapter 3.

50. The analysis of limiting cotton did not take into account long-term effects, such as existing equipment and other investments. Therefore, the analysis explicitly assumes that limiting crops is a temporary policy that is implemented only for a short period.

Chapter 6

1. There is one additional source. Israel provides 5 mcm annually to the Gaza Strip at a price of 40 cents per cubic meter.

2. Of course, the water currently used by the Israeli settlements in Gaza affects this also.

3. The exception is water consumed in parts of Gaza, where salinity levels exceed those recommended by the World Health Organization. However, this relatively saline water continues to be used in all sectors.

4. One exception is water provided by the United Nations Relief and Works Agency for Palestinian Refugees in the Near East to refugee camps, where water prices are fixed at 10 cents per cubic meter, but this water represents a small enough percentage of the total consumption that we ignore its lower cost in the analysis.

5. The agricultural sector in Gaza is not charged for water, but farmers do pay the cost of pumping. Because the cost of pumping can be highly variable, we use 10 cents per cubic meter as an average cost.

6. A more thorough treatment of unpaid-for water is given in Chapter 7, on Jordan. Note that projections for 2010 and 2020 assume that the issue of unpaid-for water has been fully addressed, and it is therefore not considered further in this chapter after the 1995 runs.

7. Note that consumption exceeded the estimated supply in the East Jerusalem district for 1995. East Jerusalem in 1995 did in fact receive water from the Israeli system. We further address this issue below.

8. The GDP growth rates are primarily derived from a World Bank (1993) report, *Developing the Occupied Territories: An Investment in Peace.* The low-scenario annual real GDP growth rates are presumed to average 5.5% for 1995–2000, 6.2% for 2000–2010, and 5.1% for 2010–2020. These growth rates rise as high as 7.2% under more optimistic projections.

9. All these estimates were made prior to the *intifada* that began in fall 2000, which undoubtedly has had a serious impact. However, given the uncertainty around this issue, we continue to use the estimates from before the war. Presumably, when peace is restored, these estimates will be approximately valid, if delayed by a few years.

10. This is defined as the derivative of the logarithm of demand with respect to the logarithm of income, or, less precisely, as the percentage change in demand brought about by a 1% change in income, other things being equal.

11. Ideally, the calculation would be made based on personal income, but since the data are not readily available, GDP data are used as an approximation.
12. Delivery of water between Israel and Palestine is explored in more detail in Chapter 8, in a discussion of cooperation.
13. In Chapter 8, we present the finding that under cooperation it would be efficient to have a recycling plant in Gaza with at least some of the output sold to Israel for agricultural use in the Negev.
14. This is without the addition of desalination plants in Gaza, an option considered below.
15. Note, however, that desalination would certainly be needed in times of considerable drought. Further, we have taken no account of the possibility that desalination might be desirable for repair of the Gazan aquifer. (We are indebted to Karen Assaf for pointing this out to us.)
16. As already discussed, the name given to this body of water—also known as the Sea of Galilee and the Kinneret—is that used in the country studied.

Chapter 7

1. Note that the long-term average availability of water can change because of changes in infrastructure to store additional water, as well as because of climate change.
2. As already discussed, the name given to this body of water—also known as the Sea of Galilee and the Kinneret—is that used in the country studied.
3. Israel once threatened to reduce this amount under drought conditions, but to date, these quantities have always been provided. In 1995 only 30 mcm was transferred from the Upper Jordan to the King Abdullah Canal. The remaining 25 mcm started to flow in May 1997.
4. All the data and projections presented can be easily changed by the user from the interface of the WAS computer program.
5. In fact, 1995 data were not readily available; data for the domestic sector are from 1996, and for the industrial and agricultural sectors, from 1998. This is assumed adequate for comparison with 1995 data for Israel and Palestine.
6. The effects of changes in elasticity can be readily explored through the WAS interface.
7. Recycled water is also used in industry, but the quantity is sufficiently small to be ignored.
8. In addition to the interdistrict system, Jordan has several conveyance systems that transport water within a district.
9. As discussed in Chapter 6, Section 3, leakage plays an important role in water management in terms of the amount of water that ultimately reaches a consumer. More subtle is the effect leakage can have on decisionmaking for infrastructure, particularly that of interdistrict pipelines. If the shadow values between any two districts differ by more than the marginal cost of conveying water between the districts, it is possible that it would be cost-effective to build a conveyance line between those two districts. The decision to construct the pipeline will be positive if the benefits obtained from the conveyed water outweigh the capital cost of construction. However, it is critical that this comparison of shadow values between districts use values before accounting for intradistrict leakage, since the greater the intradistrict leakage, the lower the shadow value of water conveyed there. A comparison of shadow values after leakage, which are necessarily

higher, would therefore be misleading because the actual gain in social welfare is that represented by the before-leakage shadow values. Therefore all shadow values shown in this chapter are before accounting for intradistrict leakage. Of course, reduction in intradistrict leakage is highly desirable and is considered in scenarios for the future in the following sections.

10. Note that there is no similar problem (for 1995) between Zarqa and Amman or Balqa and Amman. There, conveyance limits were adequate in 1995.

11. These quantities can be changed through the interface.

12. The water crisis in the projection years is already extremely severe without such policies.

13. Correspondingly, the value of the unpaid-for water quantities given in Table 7.4.2 is approximately $300 million per year in 2010 and exceeds $2 billion per year in 2020, primarily because of the water problem in these districts.

14. See Table 7.2.8.

15. Source: Ra'ed Daoud, personal communication.

16. There is a possibility that the aquifer extends much closer to Amman, only 80 kilometers, which would considerably reduce the transport cost. This possibility is still under exploration (personal communication, Munther Haddadin).

17. Note, however, that it also assumes that no other relevant infrastructure will be built. See the discussion of the proposed Red Sea–Dead Sea Canal, below, for possible implications of this type of consideration.

18. Note that in the case of desalination plants, unlike other infrastructure types, capital costs are already included in the estimate of cost per cubic meter. Hence, there is no need to calculate the discounted value of net annual benefits to compare it with capital costs.

19. See Table 7.2.8.

20. This appendix is a very slightly edited version of Salman, Al-Karableh, and Fisher (2001). It is reprinted with permission from Elsevier.

21. There is a relatively large reduction in the use of brackish water. However, the shadow price of brackish water in the model results is zero, so that the optimized value of the objective function would not be changed (at least within some range) by changes in the amount of brackish water used or its distribution over seasons.

22. Earlier uses of AGSM have also not examined the possibility of blending different water qualities. This, however, is a matter of the particular activities specified by the user and not a shortcoming of AGSM itself.

23. In this appendix, we use the term "shadow prices" instead of "shadow values" as used elsewhere in the book. This is to avoid confusion with the shadow values generated by WAS.

Chapter 8

1. There is a minor infrastructure problem regarding the Jordan Valley settlements district claimed by Israel. In the runs involving drought and with Israeli fixed-price policies in effect, that district does not have enough water to deliver the amounts demanded under those policies, and an infeasibility occurs in the results. To avoid this, in such cases, we assume a conveyance line from the Jerusalem Mountains district to the Jordan Valley settlements district, carry-

ing water at 5 cents per cubic meter. This solves the problem (with very little impact on the results for other districts). The line carries 8.4 mcm in 2010 and 5.8 mcm in 2020. A similar device is used in the "gains" runs below for noncooperative cases in which Israeli fixed-price policies are in effect and Israel is assumed to own relatively little water.

2. This is actually Yarmouk water stored interseasonally for Jordan in the Kinneret-Lake Tiberias. Of course, in this run, Jordan is selling the remainder of the 55 mcm to the other parties and gaining thereby. We explore such gains below.

3. But note that, in time of drought, Israel cuts back the amount of subsidized water allocated to agriculture. That effect is not taken into account in our results.

4. Of course, by Western standards, there is already a water crisis in Amman.

5. Note, in this regard, that the presence or absence of Israeli fixed-price policies matters to Jordan (and as seen above, the presence or absence of Jordanian fixed-price policies matters to Israel—and both matter to Palestine). We examine such matters further in the gains runs below.

6. The Mountain Aquifer actually consists of several subaquifers. It is very difficult to secure accurate information on how the water in each of these is now divided. The 76–24 split is therefore an approximation applying to the total. In the runs reported below, where necessary, we have used that split to represent existing circumstances. Of course, the general conclusions are not affected by this, and the quantitative results cannot be far off.

7. Here and later, the gains with this division are so large as to dominate the scale of the figure. This must be taken into account in examining the results.

8. Although the qualitative conclusions remain the same, the quantitative results are substantially different from those presented in Fisher et al. (2002). This is due partly to improved data, but mostly from a more realistic treatment of intradistrict leakage in Palestine, which affects the value of water.

9. This 55 mcm corresponds to the Yarmouk water stored interseasonally for Jordan in the Kinneret-Lake Tiberias.

10. The gains to Israel at the right-hand sides of Figures 8.4.1B and 8.4.2B are understated. In the cases in which Israel owns only 20% of the water and there is no cooperation, substantial additional conveyance infrastructure is required to make Israel's fixed-price policies feasible in various districts. The presence of such infrastructure has been assumed in the noncooperative runs, but the costs thereof have not been taken into account.

11. See note *a* to the table.

References

Amir, I., and F.M. Fisher. 1999. Analyzing Agricultural Demand for Water with an Optimizing Model. *Journal of Agricultural Systems* 61: 45–56.

———. 2000. Response of Near-Optimal Agricultural Production to Water Policies. *Journal of Agricultural Systems* 64: 115–30.

Amir, I., J. Puech, and J. Granier. 1991. ISFARM: An Integrated System for Farm Management, Methodology. *Journal of Agricultural Systems* 35(4): 455–69.

———. 1992. ISFARM: An Integrated System for Farm Management, Applicability. *Journal of Agricultural Systems* 41(1): 23–39.

Brown, G., and C. McGuire. 1967. A Socially Optimum Pricing Policy for a Public Water Agency. *Water Resources Research* 3(1): 33–43.

Coase, R. 1960. The Problem of Social Cost. *Journal of Law and Economics* 3: 1–44.

Dandy, G., E. McBean, and B. Hutchinson. 1984. A Model for Constrained Optimum Water Pricing and Capacity Expansion. *Water Resources Research* 20(5): 511–20.

Draper, A.J., M.W. Jenkins, K.W. Kirby, J.R. Lund, and R.E. Howitt. 2003. Economic-Engineering Optimization for California Water Management. *Journal of Water Resources Planning and Management-ASCE* 129: 155–64.

Eckstein, O. 1958. *Water-Resource Development: The Economics of Project Evaluation.* Cambridge, MA: Harvard University Press.

Eckstein, Z., and G. Fishelson. 1994. The Water System in Israel. Paper prepared for the Middle East Water Project, Tel Aviv University.

Eckstein, Z., D. Zackay, Y. Nachtom, and G. Fishelson. 1994. The Allocation of Water Resources Between Israel, the West Bank and Gaza: An Economic Analysis. *Economic Quarterly* 41: 331–69 (in Hebrew).

Fisher, F.M., S. Arlosoroff, Z. Eckstein, M. Haddadin, S. Hamati, A. Huber-Lee, A. Jarrar, A. Jayyousi, U. Shamir, and H. Wesseling. 2002. Optimal Water Management and Conflict Resolution: The Middle East Water Project. *Water Resources Research* 38(11): 25/1–25/13.

Hirshleifer, J., J.C. De Haven, and J.W. Milliman. 1960. *Water Supply: Economics, Technology, and Policy.* Chicago: University of Chicago Press.

Hotelling, H. 1931. The Economics of Exhaustible Resources. *Journal of Political Economy* 30: 137–75.

Jenkins, M.W., J.R. Lund, and R.E. Howitt. 2003. Using Economic Loss Functions To Value Urban Water Scarcity in California. *Journal of the American Water Works Association* 95: 58–70.

Jenkins, M.W., J.R. Lund, R.E. Howitt, A.J. Draper, S.M. Msangi, S.T. Tanaka, R.S. Ritzema, and G.F. Marques. 2004. Optimization of California's Water Supply System: Results and Insights. *Journal of Water Resources Planning and Management* 130: 271–80.

Jordan Valley Authority. 1997. *Review of Jordan Valley Authority Irrigation Facilities.* Amman: Ministry of Water and Irrigation.

Jordanian Ministry of Water and Irrigation. 1993. *Study for Recovery Operation and Maintenance Costs of Irrigation Water in Jordan.* Amman: Ministry of Water and Irrigation.

———. 1999. *Annual Report. Jordan Valley Authority.* Amman: Ministry of Water and Irrigation.

Kally, E. 1997. *Resource in Shortage: The Water in Israel.* Haifa, Israel: Technion (in Hebrew).

Klare, M.T. 2001. The New Geography of Conflict. *Foreign Affairs* 80(May–June): 49–61.

Kneese, A.V., and J.L. Sweeney. 1985. *Handbook of Natural Resources and Energy Economics,* Vol. II. Amsterdam: Elsevier Science Publishers B.V.

Maass, A., M.M. Hufschmidt, R. Dorfman, H.A. Thomas, and S.A. Marglin. 1962. *Design of Water Resource Systems.* Cambridge, MA: Harvard University Press.

McCarl, B. 1999. Limiting Pumping from the Edwards Aquifer: An Economic Investigation of Proposals, Water Markets and Spring Flow Guarantees. *Water Resources Research* 35(4): 257–68.

Mimi, Z., and M. Smith. 2000. Statistical Domestic Water Demand Model for the West Bank. *Water International* 25(3): 464–68.

Newlin, B.D., M.W. Jenkins, J.R. Lund, and R.E. Howitt. 2002. Southern California Water Markets: Potential and Limitations. *Journal of Water Resources Planning and Management* 128(1–2): 21–32.

Palestinian Central Bureau of Statistics (PCBS). 1994. Summary Statistics. Available at www.pcbs.org.

Rogers, P.P., and M.B. Fiering. 1986. Use of Systems Analysis in Water Management. *Water Resources Research* 22: 146S–158S.

Rostow, E.V.D. 1948. *A National Policy for the Oil Industry.* New Haven: Yale University Press.

Saliba, B.C., and D.B. Bush. 1987. *Water Markets in Theory and Practice: Market Transfers, Water Values, and Public Policy.* Boulder, CO, and London: Westview Press.

Salman, A., E. Al-Karablieh, and F.M. Fisher. 2001. An Inter-Seasonal Agricultural Water Allocation System (SAWAS). *Journal of Agricultural Systems* 68: 233–52.

Sher, A., and I. Amir. 1993. Optimization with Fuzzy Constraints in Agricultural Production Planning. *Journal of Agricultural Systems* 45: 421–41.

United Nations Environment Program (UNEP). 2002. *Global Environmental Outlook 3*. London: Earthscan.

Volk, A. 1993. Economic Data for Field Crops, Vegetables and Orchards for Agro-climate Districts in Israel. Rehovot, Israel: Hebrew University Faculty of Agriculture (in Hebrew).

Wolf, A. 1994. A Hydropolitical History of the Nile, Jordan, and Euphrates River Basins. In A.S. Biswas, ed., *International Waters of the Middle East from Euphrates–Tigris to Nile*. Oxford: Oxford University Press.

World Bank. 1993. *Developing the Occupied Territories: An Investment in Peace*. Washington, DC: World Bank.

Index

pp 66-67
70-73

PCQ? Formula:
1. per capita [illegible] vs 6.
2. per acre for cult. vs 6.
average [illegible] [illegible] dependency
Flow to = 90% net fixed quantity
% of ongoing the census

Fisher notes:

1) init hydrologic principles
2) stress formula capture graphically rather than mathematically

e.g. Ch. 3 – agric. demand 3.3.2 + 3.3.4 + 6.4.4

3) Rebut leakage + unauth use 130–137
20% 40%

4) recycling 140

5) WB Hebron–Gaza (?) 141

6) desalin. plants 144
NB: conclusions for policy for border (stop)
border

7) leaky + unauth — 165
28 28 — 171

PCQ?: NB: leakage / unauth # given for Israel!!!

8) agg both pts! but! for unauth conditions
tradeoff plot — 212

PC + agh 1) Zaslavsky & other eg Sayousi or Sarron (Google) for key status given + [illegible]

2) Fisher to cite
a) est. costs of diff. tech.
I 2)
b) discuss w/ policy benefits / diff. of e
~~need for kegs??~~

3) ? to compare existing quant. rather than vs. his model(s)
changes of replace w/ actual models,

discuss of non-H_2O measures

c) PCQ: discuss coop??